THE
INTERSYSTEM
MODEL

This book is dedicated to

Alice Evelyn Edwards Murray, RN

Ashland General Hospital Training School, Ashland, Wisconsin
Class of 1929

Alice understood early in her nursing career the importance of understanding the context in which her patients were living and adapted her nursing care to meet their specific needs.

She was able not only to base her practice on the lived experience of her patients but also to convey this philosophy of nursing to her daughters.

THE INTERSYSTEM MODEL

Integrating Theory and Practice

editors

Barbara M. Artinian, RN, PhD
Professor, School of Nursing
Azusa Pacific University, Azusa, California

Margaret M. Conger, RN, MSN, EdD
Associate Professor, Department of Nursing
Northern Arizona University, Flagstaff, Arizona

SAGE Publications
International Educational and Professional Publisher
Thousand Oaks   London   New Delhi

For information address:

 SAGE Publications, Inc.
2455 Teller Road
Thousand Oaks, California 91320
E-mail: order@sagepub.com

SAGE Publications Ltd.
6 Bonhill Street
London EC2A 4PU
United Kingdom

SAGE Publications India Pvt. Ltd.
M-32 Market
Greater Kailash I
New Delhi 110 048 India

Printed in the United States of America

Library of Congress Cataloging-in-Publication Data

Main entry under title:

The intersystem model: Integrating theory and practice / editors,
   Barbara M. Artinian and Margaret M. Conger.
       p. cm.
   Includes bibliographical references and index.
   ISBN 0-8039-5558-8 (cloth: alk. paper).—ISBN 0-8039-5559-6
   (pbk.: alk. paper)
     1. Nurse and patient.   2. Nursing—Psychological aspects.
   3. Nursing models.   I. Artinian, Barbara M.   II. Conger, Margaret M.
   RT86.3.I585   1997
   610.73—dc21   96-45795

97 98 99 00 01 02 03 10 9 8 7 6 5 4 3 2 1

Acquiring Editor:	Dan Ruth
Editorial Assistant:	Jessica Crawford
Production Assistant:	Karen Wiley
Typesetter & Designer:	Andrea D. Swanson
Indexer:	Mary Kidd
Cover Designer:	Lesa Valdez

Contents

Foreword

This book introduces a nursing model that has wide application in health care, especially in today's environment where people are concerned with the "spiritualization" and "demedicalization" of health. Azusa Pacific University is a small gem in the large system of higher education in America. Its nursing program is a bright star whose excellence is well known. What better place to see a philosophy and model of nursing evolve than in a place where spiritual values and academic excellence are given high priority?

When the School of Nursing was begun at Azusa Pacific University, the faculty searched for a nursing model that reflected the Christian philosophy of the university. None was found. The systems model developed by Chrisman and Riehl, however, had the potential to become a nursing model that would facilitate a focus on the spiritual core of human beings. Sally Jo Brown, a part-time faculty member, was enlisted to work on revising the Chrisman-Riehl model, and her product was called the Nursing Process Systems Model. During my years as dean (1979-1987), this model was further revised and improved. Many individuals contributed significantly to its growth, both faculty and students. It formed the basis for the curriculum and was used by every student to provide a framework for patient care. The unique feature of the model was its focus on the spiritual, biological, and psychosocial systems as central to every individual. It further proposed the concepts of development, stress, and system as the mediating influences for individuals, families, and communities dealing with issues of health and illness. Each student, over the years, has been an ambassador for the university through the use of the model in patient care. Patients, and indeed the health care community, may not have credited the model as a reason why Azusa Pacific students were so outstanding, but they certainly recognized and appreciated that the care they gave was special.

When Barbara Artinian joined the faculty at Azusa Pacific as a sociologist, she realized that an intersystem approach would make the model more versatile in dealing with different populations and health care settings and began to extend the model in that direction. This book is the outcome of the work we began one year at a faculty retreat, high overlooking the blue Pacific Ocean, and represents considerable intellectual work during the past 10 years. The authors are all faculty members or students who have worked extensively with the application of the model in practice and education, and their work is a tribute to all the patients, teachers, and students who have contributed to its development.

Marilynn J. Wood
Dean and Professor of Nursing
The University of Alberta

Overview of the Intersystem Model[1]

Barbara M. Artinian

Development of the Model
Metaparadigm Concepts
Discussion

The Intersystem Model describes the interactional process that takes place between the nurse and patient/client when nursing assistance is needed. In this model, the client can be an individual, a family, an institution, or a community, each with its own network of significant others. Likewise, the nurse can be an individual or a provider system in an institutional context within a network of significant others. In the interactional process, the two systems come together to form an intersystem characterized by the specific set of relations that connect them to each other. Intersystem interaction is mutually influencing and focuses on how information is communicated, how values are negotiated, and how behaviors are organized to develop a mutual plan of care that will increase the client's situational sense of coherence.

Development of the Model

Nursing Process System Model

The Intersystem Model was developed at Azusa Pacific University. It is based on the Systems-Developmental Stress Model developed by Chrisman and Riehl (1974) and was called the Nursing Process Systems Model (NPS) (Brown, 1981). The NPS model viewed the human being as a unified whole system made up of various subsystems (biological, psychosocial, and spiritual), which were viewed in terms of change over time (Brown, 1981). Eight assumptions about these concepts were identified:

1. The human being exists within a framework of development or change—life continuum of birth, growth and development, maturation, and death.
2. Change is inherent to life. The systems attempt to maintain stability or equilibrium within change.
3. At any point in time, a person's life can be viewed as a unit of interrelated biological, psychosocial, and spiritual systems that are open to the environment and subject to change.
4. At any point in time, the person's present can be seen in terms of his or her past and potential future.
5. The person's interaction with the environment takes place ultimately on the biological level. The five senses provide a mode for taking in from the environment, and various bodily functions provide a mode for giving out into the environment.
6. The human spirit is at the center of the person's being and has ultimate effects on all aspects of life. The spirit transcends time, possessing an eternal quality (Stallwood & Stoll, 1975).
7. In clinical practice, the nurse focuses on aspects of the total person systematically analyzing: (a) the client's biological, psychological, and spiritual systems; (b) interactions of interrelations of these systems; and (c) the relationship of the systems to time and environment.
8. Nursing process takes place in the present, taking into account the client's past and future (pp. 37-38).

The nurse used the nursing process to look at the client's systems in terms of stressors encountered and the degree of success of the stress response. A successful response to stress restored equilibrium to the system and was viewed as adaptation that resulted in health. The model stated that the nurse deals with the client in the present because "it is only in the present that nursing can assess and intervene" (Brown, 1981, p. 40).

Revised Model

A revised version of the NPS model was prepared for the self-study report to the National League for Nursing, Department of Baccalaureate and Higher Degree programs, in 1983. Elaine Goehner assumed the primary responsibility for the writing of this revision. No substantial changes were made in the model, but many parts were described in more detail. For example, the components of the spiritual subsystems were identified as love and related-ness, purpose and meaning, and forgiveness. The concept of system also was expanded to place the individual in the family system and the community system and to show how these systems and subsystems related to each other. Nursing process was described as being a reiterative process, with the evaluation of one nursing action serving as a reassessment to begin the process again. Development was seen to take place in a cultural context over time, with death as a new beginning. Another important change was the emphasis on the nurse as a person. The responsibility of the nurse to assess his or her own development and be a changing, growing person was seen as a means of facilitating client growth. The focus on identifying both positive and negative stressors made it a model that directed the nurse to look for nurse and patient/client strengths as well as weaknesses in identifying coping responses.

Intersystem Model

In the third revision of the NPS model, no substantial changes were made in the core concepts of the model. The interpersonal tradition of Peplau (1952), which emphasized the relationship between nurse and patient, and the deliberative nursing process of Orlando (1961), which focused on clear communication that allowed accurate assessment of needs and mutual goal setting, were continued in the development of the Intersystem Model. A new construct, situational sense of coherence (SSOC), adapted from Antonovsky's (1987) sense of coherence (SOC), was incorporated into the model to measure the client's integrative potential in his or her response to an illness situation. In addition, the process of nursing action was identified to be the intrasystem analysis and the intersystem interaction developed by Kuhn (1974). The name of the model was changed to the Intersystem Model to reflect the interactional process that takes place between client and nurse in using the nursing process to develop a plan of care. The systems approach provides a way of looking at "person" as a whole dynamic entity in a hierarchical relation of subsystems and suprasystems in mutual interaction. The major concepts of the model were placed within the four metaparadigm concepts of nursing: person, environment, health, and nursing action.

Metaparadigm Concepts

Person

The first of the metaparadigm concepts in a nursing model is the person receiving nursing care. In the Intersystem Model, a person is viewed as a coherent being who continually strives to make sense of his or her world.

Person is seen as a system made up of various subsystems that can be viewed in terms of change over time. The subsystems that make up the whole person are the biological, psychosocial, and spiritual. These are configured in a specific arrangement so that transactions among the parts result in emergent properties at the systemic level. Therefore, as stated by Whitchurch and Constantine (1993), the "system must be understood as a whole and cannot be comprehended by examining its individual parts in isolation from each other" (p. 328). This is based on the general systems theory assumption that the whole is greater than the sum of its parts. The subsystems are illustrated as a series of concentric circles with the biological subsystem on the outside, since it is through the body that the psychosocial and spiritual subsystems are manifested in the environment (Figure 1.1). The spiritual subsystem is in the center, since the spirit is viewed as being the core of the person's being. A person is seen developmentally as moving from conception to death and beyond.

Maturana and Varela describe the person as a perceiving, self-determining, and self-regulating human system who operates according to how he or she is built and put together at any moment in time (Maturana & Varela, 1987). The person is self-aware, has feeling, and a sense of who he or she is. The person can anticipate, make decisions, and act upon these decisions.

According to Maturana and Varela, every individual has an underlying core that enables him or her to integrate human experiences. This core is the niche of the person (Maturana & Varela, 1987). Maturana and Varela refer to the niche of a person as the consciousness, identity, soul, or personhood of the person. It is formed by the way in which biopsychosocial and spiritual domains of experience are organized relative to one another.

Although the term *client* had been used consistently in the first version of the NPS model, it was decided to use either the term *client* or *patient* in the Intersystem Model. The term client is derived from the Latin word meaning "to lean on" and implies that the person as client needs the professional advice or services of the nurse. The term patient is derived from the Latin word meaning "to suffer" and implies that the person as patient needs the caring the nurse can provide in a time of distress. Therefore, the choice of term will depend on the need of the patient/client in a particular interactional situation. Fawcett (1993) agrees that the term client should not replace the term patient to describe the recipient of nursing care. She writes, "Client, however, reflects a particular view of the person, and, therefore should not be used to represent phenomena of interest to the entire discipline of nursing" (p. 3).

Person as Individual

Biological Subsystem. The components of the biological subsystem are physical parts of the body that are grouped together to perform particular functions for the individual. They are:

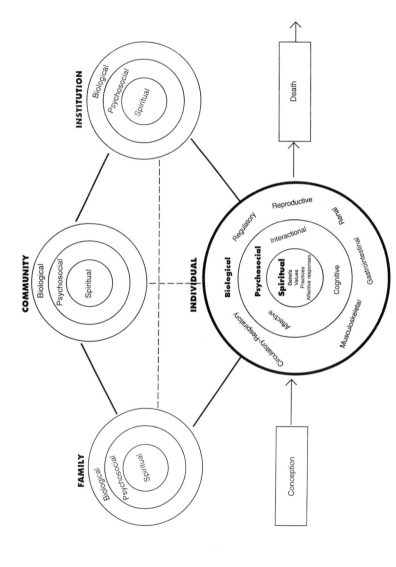

Figure 1.1. Hierarchical relationships of systems (adapted from Stallwood & Stoll, 1975). Used with permission from Blackwell.

1. regulatory subsystem:
   a. neurological,
   b. endocrine,
   c. immune;
2. circulatory-respiratory subsystem;
3. musculoskeletal subsystem;
4. gastrointestinal subsystem;
5. renal subsystem; and
6. reproductive subsystem.

For each biological subsystem, the person has knowledge about its functions, has attitudes and values, and has a repertoire of behaviors relative to it. In addition to manifesting behaviors relating to the biological subsystem, the biological subsystem is also the physical means of manifesting the cognitive, affective, and behavioral components of the spiritual and psychosocial subsystems. For the sake of understanding, it is useful to examine each subsystem separately, but in actuality the person functions as a whole.

Psychosocial Subsystem. The psychosocial subsystem of the individual is that component which enables the person to relate to self and others. In interaction with others, the person forms perceptions of self and the world. Examples of concepts that emanate from this subsystem are body image, autonomy, self-concept, role, IQ, locus of control, developmental tasks, perception of health, emotional status, attitudes, coping patterns, and motivation. The psychosocial subsystem can be divided into three subsystems:

1. cognitive;
2. affective; and
3. interactional.

The cognitive subsystem is the information-processing part of the person and is limited by the intellectual level, mental status, and perceptions of the person. The worldview of the person determines what is important for him or her. The affective subsystem displays the feelings and emotions of the person, which stem from his or her temperament. The interactional subsystem is the behavioral part of the person. It enables him or her to form a view of self in relation to others that makes possible the person's repertoire of behaviors, coping strategies, rituals, and traditional practices.

Spiritual Subsystem. The spiritual subsystem of the individual is that component which is "the life principle which pervades a person's entire being and which integrates and transcends biopsychosocial nature" (Gordon, 1985,

p. 262). The term *spiritual* is used as the descriptor for the inner center of the person from which religious beliefs and practices flow. Spiritual knowledge is organized in a statement of beliefs about deity and how to relate to deity, and produces hope by giving meaning and purpose to life. The four subsystems of the spiritual subsystem are

1. spiritual beliefs;
2. spiritual values;
3. religious practices; and
4. affective responses.

Person as Aggregate

Systems theory provides not only a logical method of looking at the subsystems of a single individual but also a framework for observing the relationship of individuals to the suprasystems of which they are a part. For example, the individual can be part of several subsystems of the family system such as parent-child subsystem, spouse subsystem, or sibling subsystem. The family is also a subsystem of the community or of institutions in the community. Systems theory is useful in describing the interrelationships among all these systems with their sub- and suprasystems. For example, we could examine the relationship of a family to a health care provider and how both the family and that provider relate to other health agencies in the community.

Because of the hierarchical nature of the systems approach, *person* can be variously defined depending on the unit being used for analysis. For example, it is possible for the client of the nurse to be the community, an institution, the family, or an individual. If the family is considered to be the client, then the community is its suprasystem and the individuals within the family are the subsystems. At the other end of the spectrum, if a nurse physiologist is concerned with angiotensin production in the renal subsystem, this systems framework helps to relate work at the cellular level with the larger system of which it is a part. These hierarchical relationships are illustrated in Figure 1.1.

The biological, psychosocial, and spiritual subsystems can be assessed for aggregates such as families, institutions, or communities using aggregate data, such as family assessment inventories, vital statistics, or surveys assessing specific characteristics of the aggregate.

Environment

Developmental Environment

Merriam-Webster's Collegiate Dictionary (1993) defines environment as "the aggregate of social and cultural conditions that influence the life of an individual or community." Kuhn expands the definition to include the bio-logical system of the individual as the internal environment because he says that the "individual learns and makes decisions about his own body by the same processes as for the world outside his skin" (Kuhn, 1974, p. 40).

Therefore, the developmental environment can be defined as all the events, factors, and influences that affect the system experienced through its subsystems (biological, psychosocial, and spiritual) as it passes through its developmental stages. In systems theory, whatever is not the system is its environment. In this way, each suprasystem or subsystem is the environment of the system. Lewin (1951) commented on the interdependency of person and environment when he said that it is not possible to understand or predict behaviors without considering them as a constellation of interdependent factors. In his general systems theory, Bertalanffy (1968) argued that, in an open system, we must observe how a system acts on its environment, how the environment acts on it, and how that interaction affects the system's growth and survival.

The developmental environment is the arena in which the system is devel-oped. It provides the context for the interaction at the time of a specific encounter in which nursing service is offered to assist with an illness or management problem the system is unable to resolve independently. Sym-bolic interaction theory helps in understanding the interaction that takes place in the developmental environment. Symbolic interactionism focuses on the nature of human social interaction by examining the connection between symbols (shared meanings) and interactions (verbal and nonverbal actions) with which individuals in relationship with each other create symbolic worlds. It is in this symbolic environment that persons develop their explanations for illness, their expression of symptoms, and their patterns for obtaining care. During the interaction of the nurse with the patient/client, the nurse is able to assess the effect of both the external and internal environments on the biological, psychosocial, and spiritual subsystems of the individual, family, institution, or community.

Situational Environment

The event that initiates interaction between the client and the nurse occurs in the situational environment. The situational environment includes all the details of the encounter such as the place, time, circumstances, motivational state, and receptivity of the care provider and recipient. The situational constraints of the particular environment determine the repertoire of behaviors that can be used and limit the outcomes available. The interaction that takes place during the encounter becomes part of the developmental environment when it is concluded. The nursing care that is provided during the situational interaction can be at the primary, secondary, tertiary, or terminal level.

Health

Health and disease are viewed on a multidimensional health/disease continuum. It would be difficult to find a person in perfect health, totally sound in body, mind, and spirit with full vigor and freedom from all signs of disease. On the other hand, as long as individuals live, they have some soundness of body, mind, and spirit.

Antonovsky (1987) suggests that people can manage stress and stay well by developing a strong sense of coherence (SOC), which he defines as

> a global orientation that expresses the extent to which one has a pervasive, enduring though dynamic feeling of confidence that (1) the stimuli deriving from one's internal and external environments in the course of living are structured, predictable, and explicable; (2) the resources are available to one to meet the demands posed by these stimuli; and (3) these demands are challenges, worthy of investment and engagement. (p. 19)

When individuals have a strong SOC, they can use a variety of coping strategies, which he defines as generalized resistance resources, in seeking a solution to the problem. If a person has a low SOC or the stressor generates more tension than can be managed, the person experiences a generalized resistance deficit and is in need of assistance to develop strategies to cope with the problem.

Antonovsky (1987) views the SOC as a "deeply rooted, stable dispositional orientation of a person" (p. 124) but states that there can be fluctuations around a mean, as in a time of crisis. In the Intersystem Model, the change in SOC brought about by a crisis event is called the situational sense of coherence (SSOC), and the goal of nursing action is to assist the person to increase in SSOC when confronted with stressors that cannot be managed independently. When a person is confronted with a stressor, the outcome can be pathological, neutral, or salutary, depending on the adequacy of the tension management used to deal with the perceived stressor. The coping responses available to the individual, as well as the potential resources, determine whether the emotional tension generated by the stressors will be transformed into a stress response that is inadequate or unsuccessful. When this happens, maladaptation occurs, which will be reflected in a low SSOC.

When a successful stress response occurs, the response is adaptive and is reflected in a higher SSOC. Kuhn (1974) defines adaptation as behavior that "changes the relation of a system to its environment" so that the "likelihood of achieving some goal of the system is increased" (p. 38). Adaptation moves the person toward a higher SSOC. This conceptualization of adaptation is similar to the "adaptation level" as defined by Roy and Andrews (1991).

By assisting the client to increase the level of SSOC, the nurse helps to ensure that the client will be able to achieve the best possible state of health or outcome given the circumstances. Health is a dynamic state of functioning within the limitations of the person. Through strengthening a person's SSOC, health is achieved by successful adaptation to the stressors in the internal and external environment. Therefore, in the Intersystem Model, health is defined as a strong SSOC. This means that the person has confidence that events are comprehensible, worth investing in, and manageable.

Nursing Action

To assist the client to increase in SSOC, the goals of both the nurse and the patient/client must be met. The nurse as a professional person has knowledge, values, and behavioral skills that help him or her identify tentative goals for the person.

This is done by assessing the client's knowledge about the problem (comprehensibility), the resources available to make it manageable (manageability), and the client's motivation to accept the challenges created by the problem (meaningfulness). This is scored as the client's SSOC, on a scale of high, medium, or low. However, these tentative goals cannot be achieved for the person until the nurse understands the life situation of the person and comes to understand in a deep way what his or her goals are. Thus, as nurse and patient communicate with each other, the meanings, expectations, and understandings that each brings to the situation can be explored. This makes it possible for the nurse to assist the patient to formulate personal goals. A systematic way to explore the definitions of the situation held by patient and nurse, to develop a plan of care that is acceptable to both interactants, has been developed using the social systems model of Alfred Kuhn (Artinian, 1983, 1991).

The Kuhn Model

The Kuhn model (1974) consists of two interrelated models: the intrasystem model and the intersystem model. Kuhn states that any controlled adaptive system must use knowledge, preferences or values, and behavioral responses. In the intrasystem model, these variables are handled respectively by the detector function, which supplies information about the environment; the selector function, which analyzes internal values and preferences; and the effector function, which initiates action in keeping with the goals of the system (pp. 43-44).

The Intrasystem Model. The intrasystem model is used to clarify the understanding of both the client and the nurse individually. The detector part of the system provides information about the state of a system's environment, including physical symptoms, knowledge base, social situations, treatment, and resources. The selector part of the system compares the state of the system with goals of the system and selects a response from the repertoire of behaviors available. If a person experiencing an illness problem has the coping resources that enable him or her to display a strong SSOC about the problem without nursing intervention, the problem would not come to the attention of the nurse. On the other hand, if the client experiences a coping deficit that can be remedied by nursing assistance, a nurse-client interaction is initiated. Nursing assistance is always needed when a client enters the hospital environment, because even a client with a strong SOC needs assistance in managing hospital protocols.

The Intersystem Model. The second model, the intersystem model, is an interpersonal model that describes the interaction that takes place when nursing assistance is needed. An intersystem model consists of two intrasystem models that are connected to each other through a specific set of relations. An intersystem model differs from a systems model with subsystems in that each component of the intersystem model retains its autonomy. Chin (1980) says, "The intersystem model exaggerates the virtue of autonomy and the limited nature of interdependence of the interactions between the two connected systems" (p. 29). He describes the advantages of an intersystem model as follows: (a) The external change agent does not completely become a part of the client system; (b) intersystem analysis of the change agent's role leads to a fruitful analysis of the connectives: their nature in the beginning, how they shift, and how they are cut off; (c) intersystem analysis allows study of an unexplored issue, the internal system of the change agent; (d) relational issues are more available for diagnosis (the model can be applied to problems of leadership, power, communication, and conflict); (e) it leads to an examination of the interdependent dynamics of interaction, both within and between the units; and (f) the essence of collaborative planning is contained in an intersystem model (pp. 28-29). When two intrasystems interact, the focus is on how information is communicated between the two systems, how values are negotiated, and how behaviors are organized to develop a joint plan of care (Artinian, 1983, 1991).

The relations between the components of the intrasystem/intersystem model developed by Kuhn and the nursing process are shown in Figure 1.2. Because the Intersystem Model allows assessment of both the individual intrasystems and the interaction between the systems, it provides a systematic way to assess the initial SSOC of the client about the problem and provides a framework for helping to increase it. As client and nurse work together to develop successful

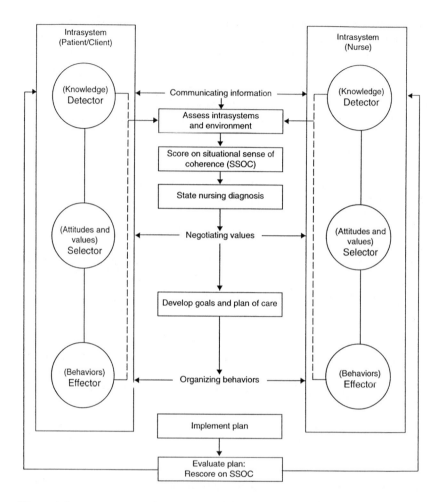

Figure 1.2. Intersystem Model (adapted from Kuhn, 1974). Used with permission from Jossey-Bass.

coping strategies, both increase in their ability to manage this problem and face new problems in the future. If the client and the nurse are not able to agree on a plan of care, the nurse will need to begin another round of communication and negotiations in an effort to develop a plan of care that is mutually acceptable. When either the emotional or physical safety of the client is in question, the nurse must exercise professional power to protect the client, recognizing that this action constitutes a failure of the interactional process and client growth should not be expected at this time.

For the nursing process to be effective, the priorities of both intrasystems must be communicated, negotiated, and organized to develop the plan of care. Through feedback loops, the intrasystem's detector, selector, and effector functions for both client and nurse are progressively clarified and modified and brought closer to a mutual plan of care to ameliorate the problem. This process is important for the development of both the client and the nurse. The same process occurs when the client is an aggregate. Just as input into the individual system is a life event that creates change, input into the aggregate is also a change event that is processed through the biological, psychosocial, and spiritual subsystems of the aggregate whether defined as family, institution, or community. Outcome is expressed as the knowledge, values, and behaviors of the aggregate, which are scored as SSOC. This constitutes the assessment part of the model. In the intervention part of the model, input is the nurse-client interaction. The purpose of the intervention is to change SSOC when it is judged to be low. This is done by changing characteristics of the subsystems or the processing of the event in the subsystems through mutual goal setting (throughput). Outcome is scored as change in SSOC by assessing changes in knowledge, values, beliefs, and behaviors.

Implementation of the Model

The Intersystem Model has been used at Azusa Pacific University by both undergraduate and graduate students. To clarify the steps to follow in using the model, a flowchart was developed (see Figure 1.3). The assessment of the effect of the developmental environment on the biological, psychosocial, and spiritual subsystems of both the client and the nurse and an analysis of the intrasystem processing of the client situation by both client and nurse provide a database for formulating a tentative diagnosis so as to begin interaction at the intersystem level. Depending on the length of time available, the initial assessment may focus primarily on only the subsystem identified as problematic for the client in the situational environment. Minimal data may be collected about the other subsystems and the developmental environment. Additional data can correct errors in the initial diagnosis as more interactions take place. To begin a database for making a nursing diagnosis, questions such as those detailed in Table 1.1 can be asked.

The next step is to analyze the stressors and coping resources that positively or negatively affect the client's SSOC, that is, variables that affect the client's understanding of the situation, the willingness to work on the challenges of the situation, and the resources available to meet the demands posed by the situation. A score can be assigned to each component of the SSOC (low = 1, medium = 2, high = 3). Each score is justified through anecdotal statements.

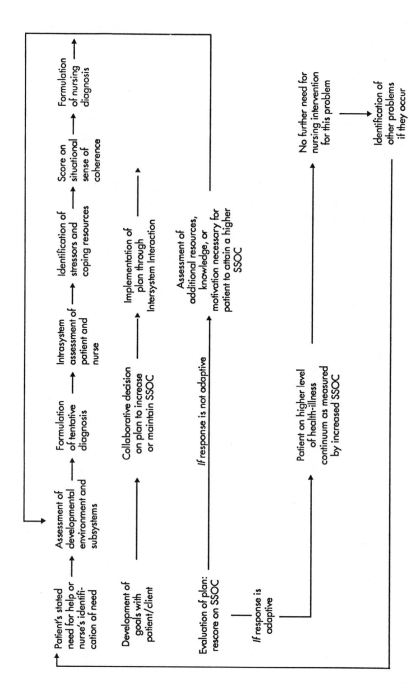

Figure 1.3. Summary of the clinical decision-making process used in implementing the Intersystem Model (adapted from Artinian, 1991). Used with permission from Blackwell.

Table 1.1   Intrasystem Assessment

Patient/Client	*Nurse*
Detector	Detector
What does client know about physical symptoms, treatment, prognosis, or resources available for management of the presenting problem?	How much information does the nurse have about presenting problem and the cultural, developmental, or role characteristics of the client and his or her significant networks?
Selector	Selector
What attitudes, values, and beliefs does the client have that may influence a plan of care or provide strength to face presenting problem?	What biases does the nurse have that may affect the plan of care? What attitudes, values, and beliefs does the nurse have that provide strength to assist the client?
Effector	Effector
What coping strategies, role relationships, technical skills, or religious practices does the client have that will influence the carrying out of plan of care?	What behaviors does the nurse manifest that may affect the interrelation, such as fatigue, shyness, technical skills, religious beliefs, supportive role relationships?

After identifying those components of the SSOC on which the client has scored low, a nursing diagnosis can be made. By negotiating values, goals can be developed and a plan can be decided on. By organizing behaviors, the plan can be implemented with the goal of increasing or maintaining the client's SSOC. Evaluation of the plan is done by again scoring the client on level of SSOC. An increased SSOC indicates that the intervention assisted the client to use coping resources to adaptively manage the stressors in the internal and external environment and thus achieve a higher state of health. The client may be discharged unless other problems are identified. If, however, the client's response is judged to be maladaptive, then additional assessment is needed to formulate a new plan.

Discussion

The Intersystem Model has been used in a variety of clinical settings including community health nursing (Artinian, 1983) and prenatal, labor and delivery, neonatal (Artinian, 1984), and acute care settings. A research study has been done by Esslinger to assess the reliability of the scoring of SSOC by students who are using the Intersystem Model. Research has also been done by Artinian to modify the orientation-to-life questions measuring SOC that were developed by Antonovsky. This research has focused on identifying which of the 29 items in the original scale are of most importance in assessing SSOC associated with a particular illness experience. Both of these studies are reported in Chapter 2.

There is need for a model that focuses on the interaction between nurse and client and specifically requires that the nurse assess the knowledge base, the values, and the behaviors that are brought to a specific client problem by both the nurse and the client. In a review of nursing research studies done within the four domains of nursing knowledge (client, client-nurse, practice, and environment), Kim (1989) found that the fewest number of research studies of any type focused on client-nurse interactions. Research based on the Intersystem Model focuses on how nurses perceive the world of their clients and guide their interactions based on that knowledge.

The Intersystem Model can be used in brief nurse-client encounters or in long-term interactions. Each time the feedback loop is completed, the nurse and client will know more about each other, and the plan they organize can be more appropriate to the client's needs. The Intersystem Model can be used by the novice practitioner as well as by the expert because the complexity of the model derives from the knowledge base of the user, not from the structure of the model.

Note

1. From Artinian (1991). The development of the Intersystem Model. *Journal of Advanced Nursing, 16,* 164-205. Adapted with permission from Blackwell.

References

Antonovsky, A. (1987). *Unraveling the mystery of health: How people manage stress and stay well.* San Francisco: Jossey-Bass.

Artinian, B. M. (1983). Implementation of the intersystem patient-care model in clinical practice. *Journal of Advanced Nursing, 8,* 117-124.

Artinian, B. M. (1984). Collaborative planning of patient care in the prenatal, labor and delivery, and neonatal settings. *Journal of Obstetric, Gynecological and Neonatal Nursing, 13,* 105-110.

Artinian, B. M. (1991). The development of the intersystem model. *Journal of Advanced Nursing, 16,* 164-205.

Bertalanffy, L. (1968). *General systems theory.* New York: George Braziller.

Brown, S. J. (1981). The nursing process systems model. *Journal of Nursing Education, 20,* 36-40.

Chin, R. (1980). The utility of system models and developmental models. In J. Riehl & C. Roy (Eds.), *Conceptual models for nursing practice* (pp. 21-37). New York: Appleton-Century-Crofts.

Chrisman, M., & Riehl, J. (1974). The systems-developmental stress model. In J. Riehl & C. Roy (Eds.), *Conceptual models for nursing practice* (pp. 247-266). New York: Appleton-Century-Crofts.

Fawcett, J. (1993). *Analysis and evaluation of nursing theories.* Philadelphia: F. A. Davis.

Gordon, M. (1985). *Manual of nursing diagnosis: 1984-85.* New York: McGraw-Hill.

Kim, H. (1989, September). *Rethinking the typology of nursing research.* Paper presented at the Biennial Meeting of the Council of Nurse Researchers, American Nurses' Association, Chicago.

Kuhn, A. (1974). *The logic of social systems: A unified, deductive, system-based approach to social science.* San Francisco: Jossey-Bass.

Lewin, K. (1951). *Field theory in social science.* New York: Harper & Row.

Maturana, H. R., & Varela, F. J. (1987). *The tree of knowledge: The biological roots of human understanding.* Boston: New Science Library.

Merriam-Webster's collegiate dictionary (10th ed.). (1993). Springfield, MA: Merriam-Webster.

Orlando, I. J. (1961). *The dynamic nurse patient relationship.* New York: G. P. Putnam.

Peplau, H. E. (1952). *Interpersonal relations in nursing.* New York: G. P. Putnam.

Roy, C., & Andrews, H. A. (1991). *The Roy adaptation model: The definitive statement. Norwalk, CT: Appleton & Lange.*

Stallwood, J., & Stoll, R. (1975). Spiritual dimensions of nursing practice, Part C. In I. L. Beland & J. Passos (Eds.), *Clinical nursing* (3rd ed., pp. 1086-1098). New York: Macmillan.

Whitchurch, G., & Constantine, H. (1993). Systems theory. In P. Boss, W. Doherty, R. La Rossa, W. Schumm, & S. Steinmetz (Eds.), *Sourcebook of family theories and methods* (pp. 325-352). New York: Plenum.

2

Situational Sense of Coherence

Development and Measurement of the Construct

Barbara M. Artinian

Theories of Psychosocial Health
Sense of Coherence
Situational Sense of Coherence

In an attempt to answer the question "What promotes health?" Antonovsky (1987) developed the sense of coherence (SOC) construct. It was designed to predict and explain movement toward the healthy end of the "health-ease/dis-ease continuum" (p. 3). The SOC construct "refers to an integrated way of looking at the world in which one lives" (Antonovsky, 1993a, p. 117). It allows adaptability to whatever life has in store for the person. SOC describes a "salutogenic" orientation to life that makes possible successful coping by enabling individuals to learn to use their own resources to their best advantage when dealing with life's challenges. It is the opposite of a pathogenic orientation, which focuses on the disease process. It produces an inclination to have a particular set of behavioral responses during time of stress. The construct SOC is incorporated within the framework of systems theory, which rejects linear causality and sees the development of SOC as being interactive with health in a recursive pattern. Antonovsky (1987) states that it "leads one to think in terms of factors promoting movement toward the healthy end of the continuum" (p. 6).

Theories of Psychosocial Health

Antonovsky is not alone in developing theory about psychosocial health. Budd (1993) has developed a concept she labels "self-coherence." She makes the assumption that "humans as open systems, continually strive toward order and self-differentiation by acquiring knowledge, structuring cognitive schemata, and choosing strategies to process environmental stimuli" (p. 361). Budd views self-coherence as

> a cognitive structure used during the process of perception or interaction with the environment . . . to facilitate coping, bringing to the present experience self-awareness of one's responses to past experiences with the environment, and current motivations and goals, thereby reducing tension. (p. 361)

Kohn and Slomczynski (1990) have taken another approach in measuring people's sense of personal efficacy and their feelings of psychic comfort or pain. They have identified two principal underlying dimensions of orientation to self and society: self-directedness of orientation versus conformity to external authority and sense of distress versus a sense of well-being. They state that "self-directedness of orientation implies the beliefs that one has the personal capacity to take responsibility for one's actions and that society is so constituted as to make self direction possible" (p. 85).

Sense of Coherence

Description of the Construct

Although both the constructs of self-coherence and self-directedness of orientation have much in common with SOC, they appear to be measuring responses at a level of psychological functioning. In contrast, Antonovsky (1987) defines SOC in a wider context as

> a global orientation that expresses the extent to which one has a pervasive, enduring though dynamic feeling of confidence that 1) the stimuli deriving from one's internal and external environments in the course of living are structured, predictable, and explicable; 2) the resources are available to one to meet the demands posed by these stimuli; and 3) these demands are challenges, worthy of investment and engagement. (p. 19)

The three components of SOC are comprehensibility, meaningfulness, and manageability. These are described in the following definitions:

Comprehensibility—the extent to which one perceives the stimuli that confront one, deriving from the internal and external environments, as making cognitive sense, as information that is ordered, consistent, structured, and clear, rather than as noise—chaotic, disordered, random, accidental, inexplicable. The person high on the sense of comprehensibility expects that stimuli he or she will encounter in the future will be predictable or, at the very least, when they do come as surprises, that they will be orderable and explicable. It is important to note that nothing is implied about the desirability of stimuli. Death, war, and failure can occur, but such a person can make sense of them.

Meaningfulness—the extent to which one feels that life makes sense emotionally, that at least some of the problems and demands posed by living are worth investing energy in, are worthy of commitment and engagement, and are challenges that are "welcome" rather than burdens that one would much rather do without. This does not mean that someone high on meaningfulness is happy about the death of a loved one, the need to undergo a serious operation, or being fired, but rather that when these unhappy experiences are imposed on such a person, he or she will willingly take up the challenge, will be determined to seek meaning in it, and will do his or her best to overcome it with dignity.

Manageability—the extent to which one perceives that resources are at one's disposal that are adequate to meet the demands posed by the stimuli that bombard one. "At one's disposal" may refer to resources under one's own control or to resources controlled by legitimate others—one's spouse, friend, colleagues, God, history, the party leader, or a physician—who one feels one can count on and whom one trusts. To the extent that one has a high sense of manageability, one will not feel victimized by events or feel that life treats one unfairly. Untoward things do happen in life, but when they do occur one will be able to cope and not grieve endlessly (Antonovsky, 1987, pp. 16-18).

Antonovsky (1987) states that the SOC construct is the core of a theory of successful coping. In this salutogenic perspective, it is hypothesized that a person with a high SOC will use personal resources and the resources of others in the best possible way in dealing the challenges of life. This is done by seeing the broader picture when defining a problem and searching for coping resources, taking into account cognitive, affective, and instrumental issues simultaneously. Using this approach, both the client and the health professional are encouraged to look for salutary factors in the situation—the negentropic or order-promoting forces—to facilitate coping.

Antonovsky (1993a) describes SOC as a stable underlying personality characteristic that is rooted in social structure and culture, referred to as the developmental environment in the Intersystem Model. The generalized SOC develops within the totality of one's life experiences. Antonovsky postulates that a person's SOC is stabilized by the end of young adulthood, thereafter showing only minor fluctuations, "barring major and lasting changes in one's life situation" (p. 118).

Although the doctrine of personality stability has been widely accepted since William James claimed in 1887 that by the age of 30 the character has been set in plaster, that doctrine is now being challenged. Brim and Kagan (1980) have described the changes that can occur in personality in adulthood in their book *Constancy and Change in Human Development*. Brim has found that certain aspects of personality, such as introversion-extraversion, depression, and anxiety, remain stable, but other parts of personality, such as self-esteem, sense of control over one's life, and ultimate values, undergo the most important changes over the course of a life (Rubin, 1981). These aspects of personality identified as having potential for change are important elements in the SOC components. Another area of research that documents change in personality is crisis research. Research in the area of crisis resolution has demonstrated that the adaptation level of people can change, as illustrated in Hill's (1949) angle-of-recovery concept in his crisis model. Hill's crisis model, the ABC-X Model, defines crisis as a period of disorganization following a stressor (A). He argues that if the family has sufficient resources (B) and a favorable definition of the stressor event (C), the family would be able to reverse the disorganization and begin to reorganize with the possibility of reaching a new level of reorganization. He diagrams the process as a sharp decline, the crisis response (X), from the organization of the steady state following the stressor event followed by an angle of recovery at some level—either lower than or exceeding the level the family experienced prior to the crisis (see Figure 2.1).

Even Antonovsky (1993b) suggests that some changes may be possible. He says that there are "minor fluctuations [in SOC] in response to a particular experience" and in response to a given mode of therapy there may be "small-scale, incremental change in the lives of people which lead to small-scale but significant and meaningful change in their SOC" (pp. 15-16). In fact, he has observed that the hospital is an institution that destroys a person's SOC because it provides no opportunities for choice and disregards the person's roles.

Measurement of the Construct

Antonovsky developed a closed questionnaire to measure the sense of coherence to statistically test the relationship of SOC to health status. He initially interviewed 51 persons who had experienced severe trauma such as disability, loss of loved ones, or concentration camp internment. The transcribed interviews were independently scored by four investigators to classify the subjects as being strong or weak on SOC on a 10-point scale. Through examination of the

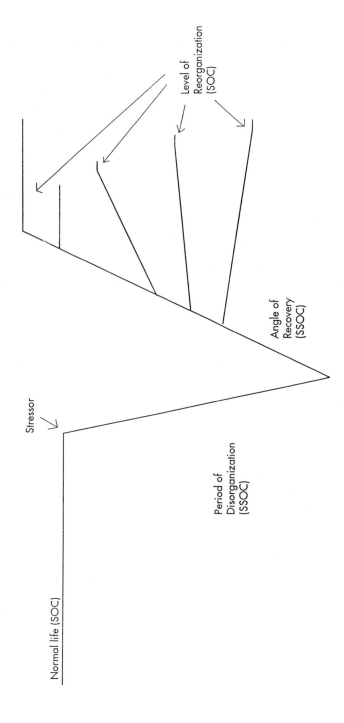

Figure 2.1. Relationship of Hill's (1949) ABC-X Crisis Model to SOC and SSOC

responses of those at the top and bottom of the scale, phrases and expressions emerged that were useful in developing the questionnaire items. Because he wanted the instrument to measure a global orientation to life, he used a wide variety of situations in developing the items. He used the technique of facet design (developed by Guttman to design mapping sentences) to develop his items (Shye, 1978). The result was a 29-item questionnaire that was field tested in 1983. It has since been tested in many studies and has shown a high level of reliability and validity (Antonovsky, 1987).

Situational Sense of Coherence

Because the global SOC construct is conceived to be stable over the life span, a new construct, the situational sense of coherence (SSOC), has been developed (Artinian, 1991). This narrower construct describes the response that occurs in the period of time in which a client is attempting to deal with a serious life event. The relative strength or weakness of the SOC influences how the person will respond to a life stressor and is reflected in the measurement of SSOC. The SSOC measures the integrative potential in the person's understanding of his situation, his or her way of looking at the situation, and the ability to marshal resources.

Description of the Construct

The SSOC contains the same three dimensions identified in the SOC, but they are defined by Artinian (1991) to reflect a present, specific orientation rather than a global orientation. The definitions are as follows:

Comprehensibility—the extent to which one perceives the stimuli present in the situational environment deriving from the internal and external environments as making cognitive sense, as information that is ordered, consistent, structured, and clear rather than disordered, random, or inexplicable.

Meaningfulness—the extent to which one feels that the problems and demands posed by the situation are worth investing energy in, are worthy of commitment, and are challenges for which meaning or purpose is sought, rather than viewed as burdens.

Manageability—the extent to which one perceives that resources at one's disposal are adequate to meet the demands posed by the stimuli present in the situation so that one does not feel victimized or treated unfairly.

The relationship of SOC to SSOC can be equated to Hill's (1949) crisis model. The straight line of normal life in the Hill Model is related to SOC as the downward line of crisis and the angle of recovery are related to SSOC. The SSOC is a measure of the client's response to the mutual plan of care. The final level of reorganization can be equated to a new level of SOC (see Figure 2.1).

The SSOC construct as used in the Intersystem Model identifies cognitions and behavioral responses to a specific illness situation. In the context of family theory, Burr (1991) states that processes that are low in abstraction—that are specific, observable, and concrete—are more amenable to change. Change can occur by introducing new coping strategies or presenting new information. Because SSOC is also a first-level construct that is more observable and concrete, it is hypothesized that change can be brought about by introducing new techniques or presenting new information, thus changing the SSOC. Burr cautions that change introduced at this level may not generalize to new situations because "Level-I processes are so situation-specific and superficial" (p. 450) that integration of the learning may not be integrated into Level II processes. At crisis points when normal life patterns are disrupted, however, individuals, families, institutions, or communities may be more open to suggestions for behavioral changes that may translate into Level II processes.

In contrast, Level II processes are more reflective and more difficult to change. In an unpublished manuscript, Antonovsky (1993b) states that "even harsh, unanticipated, uncontrollable experiences, whatever pain they cause, are likely to be integrated" (p. 15). He also states that "for any given person, a chance encounter, a courageous decision, or even an externally imposed change may initiate a considerable transformation of the level of the SOC in either direction. These are not likely" (Antonovsky, 1987, p. 123). When change at this level does occur, it results in change in a more fundamental aspect of the system. SOC, as a global orientation to life, can be considered to be a Level II process and therefore less amenable to change.

Level III processes are even more abstract and represent an individual's or group's paradigm or worldview. They are embedded in a person's culture, which can be defined as "a set of cognitive and evaluative beliefs—beliefs about what is or what ought to be—that are shared by the members of a social system and transmitted to new members" (Kohn & Slomczynski, 1990, p. 2). At the family level, the family paradigm is described by Reiss (1981) as the enduring, fundamental, and shared assumptions families create about the nature and meaning of life, what is important, and how to cope with the world in which they live (pp. 2-3). Burr (1991) states that much of Reiss's work is a demonstration that paradigms change slowly and rarely and that change would be extremely difficult to bring about. Burr also states, however, that change is not necessary at this level because this is not where the problems occur. Therefore, he suggests that practitioners can be most helpful to families if they help them adapt their Level II processes so they are consistent with the family paradigms. Burr writes, "To try to help families change their paradigmatic beliefs would usually be unwise because such beliefs usually are internally consistent . . . and are almost never pathological" (p. 450). He states that if a practitioner would try to "reframe a problem situation in a way that is inconsistent with a family's paradigmatic beliefs, it would introduce new problems rather than providing helpful ways of thinking" (p. 449).

Burr (1991) is referring to the intensive therapy sessions that are known as "family therapy," in which the client has entered into therapy for an identified problem. Therefore, the client may be willing to work on Level II changes. This is not the usual situation in the nurse-client encounter. The client has come for assistance with a specific health-illness problem. Therefore, the role of the nurse is to introduce behavioral changes at Level I that are focused on increasing the person's SSOC. It is hoped that there may be a learning effect as the client reflects on the successful adaptation to a particular situation and this will translate into Level II changes that will enable the person to more successfully respond to challenges in the future.

Measurement of the Construct

Nurse-Scored Instrument

The construct SSOC has been incorporated into the Intersystem Model as the way to evaluate a client's response to an illness situation. Within the Intersystem Model, SSOC is scored by the nurse and the score is validated anecdotally. During the interaction between nurse and client, the nurse assesses the ability of the client to integrate aspects of the illness situation that affect coping. The SSOC can be scored as high, medium, or low on each of the components of comprehensibility, meaningfulness, and manageability. The actual score assigned would be substantiated by an observational note. If it is found that nursing intervention is needed as indicated by a low score on any of the components, client and nurse work together to develop a mutual plan of care designed to assist the client to achieve an optimum level of health in the given situation. The response to the plan of care is evaluated by rescoring on SSOC when care is terminated.

Because the scoring of SSOC as high, medium, or low might appear to be arbitrary and very subjective, a study was done by Esslinger (1994) to test the reliability of scoring of SSOC by students at Azusa Pacific University. This was done as part of a larger study that examined other aspects of the SSOC construct. Twelve faculty members and 56 students participated in the study. All subjects were familiar with the Intersystem Model and the definitions of SSOC, and they regularly scored patients on SSOC as part of their clinical practice. Fifty-six hospitalized patients were assessed independently by a student and a faculty member, and SSOC scores were assigned to each patient both as a total score and as scores on each of the three components. Each student assessed only 1 patient, but the faculty members assessed all the patients of students under their supervision. Forms for data collection were coded so that the faculty and student forms for each patient could be matched for data analysis. Pearson's *r* was used to measure interrater reliability. This was found to be extremely high. The interrater reliability for the overall SSOC score was $r = .9501$ ($p = .000$). For the individual components, all interrater reliabilities were significant: comprehensibility, $r = .9199$; meaningfulness, $r = .9727$; manageability, $r = .9306$ (Esslinger, 1994). These correlations indicate significant reliability in the scoring of SSOC as used in the Intersystem Model and point to the presence of an underlying conceptualization of the SSOC construct in the minds of the students and faculty members.

Self-Report Instrument

Within the framework of the Intersystem Model, SSOC has been scored only by the nurse providing care for the client. The original instrument developed by Antonovsky to measure SOC, however, was a self-report scale. Therefore, it was considered desirable to develop a new instrument to measure the construct SSOC in a self-report format. This new instrument was developed to measure the specific responses of the client to an illness situation, the first level of abstraction. In this respect, the SSOC instrument is developed at the same level of abstraction as the family sense of coherence instrument developed by Antonovsky and Sourani (1988), who designed their instrument to be situation specific so that family members would be responding to the same situation. Although SSOC is a response to a specific life situation rather than a global orientation to life as is the SOC, the instrument used to measure it is nonspecific in that it can be used to measure the response to any life situation. In this respect, it differs from an instrument such as a self-efficacy expectation scale that must be specifically developed for each client situation. The process through which the new instrument was developed and tested is described below.

The situational sense of coherence questionnaire is a new instrument developed to enable patients to self-report their responses to an illness situation (see Figure 2.2). Items were adapted from the 29-item SOC instrument developed by Antonovsky (1987) to create this new instrument to measure SSOC. The first step in developing the instrument was to delete or rephrase all items referring to past or future judgments because SSOC is concerned with the present. The remaining items were reworded to make the situational aspect of the item apparent. This created an instrument with 15 items. Content analysis of these items was done by two nurses familiar with both the SOC construct and the SSOC construct. They analyzed each item to

Name_____ Date_____

SITUATIONAL SENSE OF COHERENCE QUESTIONNAIRE
Adapted from Antonovsky (1987) by Artinian (1993)

Here is a series of questions relating to various aspects of the situation you are experiencing. Each question has seven possible answers. Please mark the number which expresses your answer, with numbers 1 and 7 being the extreme answers. If the words under 1 are right for you, circle 1; if the words under 7 are right for you, circle 7. If you feel you are somewhere in between, circle the number which best expresses your feeling. Please give only one answer to each question.

1. In this situation, if you have to do something that depends upon the cooperation of others, do you have the feeling that it:
    1   2   3   4   5   6   7
    surely won't get done   surely will get done

2. Do you have the feeling that you don't really care about what goes on around you:
    1   2   3   4   5   6   7
    very seldom or never   very often

3. People you count on often disappoint you:
    1   2   3   4   5   6   7
    never happens   always happens

4. In this situation you have:
    1   2   3   4   5   6   7
    no clear goals or purpose   very clear goals and purpose

5. In the situation you are in, do you have the feeling that you're being treated unfairly:
    1   2   3   4   5   6   7
    very often   very seldom or never

6. Do you have the feeling that you are in an unfamiliar situation and you don't know what to do:
    1   2   3   4   5   6   7
    very often   very seldom or never

7. When you think about the situation you are in, you very often:
    1   2   3   4   5   6   7
    feel how good it is to be alive   ask yourself why you exist at all

8. When you think about what has happened to you, do you tend to:
    1   2   3   4   5   6   7
    keep worrying about it   say "ok, that's that,
                                             I have to live with it"

9. In the situation you are in, do you feel that:
    1   2   3   4   5   6   7
    you can find a solution   there is no solution

10. Do you feel that your feelings and ideas are mixed-up:
    1   2   3   4   5   6   7
    very often   very seldom or never

11. In this situation you are experiencing, do you find that you have:
    1   2   3   4   5   6   7
    overestimated or under-   seen things in
    estimated its importance   the right perspective

12. How often do you have the feeling that there's little meaning in the things you do in your daily activities:
    1   2   3   4   5   6   7
    always have this feeling   never have this feeling

Figure 2.2. SSOC Questionnaire

determine if the intent of the item in the SOC instrument had been retained. In addition, the items from the SOC instrument that were not included were assessed to determine if they were (a) essential to include, (b) all right to leave out because they were not relevant to the construct as defined for SSOC, or (c) redundant. After this review, 1 item was eliminated from the instrument and an additional 4 items were added, making a total of 18 items with 6 items for meaningfulness, 5 items for manageability, and 7 items for comprehensibility. Because Antonovsky considers the scale as a totality, he states it is not necessary to have equal numbers of items for each component.

Using the criteria established by Antonovsky (1993a) that a good item would be rated consistently as a comprehensibility, meaningfulness, or manageability item, this revised form was then analyzed by seven faculty members at Azusa Pacific University using the same procedure. The following instructions were given to them:

1. Examine all items in the 18-item SSOC instrument and the additional items from the 29-item instrument that were not included in the instrument to determine the following:
   a. The category of the item: comprehensibility, meaningfulness, or manageability.
   b. Whether the item is essential or not essential because it is inconsistent with the SSOC definition or not essential because it is redundant. Give your reason for your decision for each item.
2. Analyze the definitions of the SSOC components to determine their consistency with the definition of SOC developed by Antonovsky. Make any changes you feel are necessary.

No changes in items were suggested from this analysis. Following this review, a research associate in the research department at Barlow Respiratory Hospital reviewed the instrument for understandability by elderly patients. On the basis of her suggestions, an alternate form of the instrument was developed for patients with limited reading skills using the visual self-anchoring technique.

After discussing the 18-item instrument with other researchers who had used the 29-item SOC instrument, it was decided to further reduce the number of items because they reported that the elderly had difficulty in answering a long instrument. Twenty-one graduate students at Azusa Pacific University were asked to classify each of the items as a comprehensibility, meaningfulness, or manageability item and to rank all of them in terms of salience. Those items that were correctly classified by at least 50% of the group and were judged to be most salient were retained. This resulted in a 10-item instrument. An additional 11 senior undergraduate students who had used the SSOC construct throughout their nursing program were asked to classify the items. A comparison of their classifications with those of the graduate students confirmed the original selections and identified an additional 3 items that had good correspondence. This resulted in a 13-item scale with 4 items each for comprehensibility and manageability and 5 for meaningfulness. This 13-item form was prepared in both the Likert-type and visual analog forms.

Further testing of the scale was done in a small pilot study with patients at Barlow Respiratory Hospital to see which form was preferred. It was found that the visual analog form was easier for the elderly to complete. It was also found that one item was misinterpreted by the patients, and it was removed. This resulted in a 12-item instrument with 5 items measuring manageability, 4 items measuring meaningfulness, and 3 items measuring comprehensibility. A Likert form of the instrument has also been developed for nurses to use in scoring patients on SSOC.

The original 29-item instrument developed by Antonovsky (1987) contained a subset of 13 items that could be used as a short form to measure the SOC construct. A final review of the 12-item SSOC instrument was a comparison of the items included with the subset of 13 items from the original 29-item instrument that Antonovsky had indicated could be used as a short form to measure the SOC construct. It was found that 8 of the items identified by Antonovsky as being essential to the construct had been included. Two items were not included in the SSOC instrument because they had a past orientation (Items 5 and 25), and 2 items (Items 21 and 29) were not included because they were not consistent with the definitions of the SSOC components in that they referred to control of feelings as a manageability response. In addition, Item 16 was not included because patients found it difficult to answer during a hospital experience because it referred to normal daily activities.

Further testing of the 12-item SSOC instrument is being done at Azusa Pacific University to determine the correspondence of a patient's scoring on this new self-report instrument with the score assigned by the nurse in the evaluation of the patient's situational response using the subjective high, medium, and low categories substantiated anecdotally. The study has been approved by the institutional review boards at three hospitals used for student clinical experiences. Patients for whom undergraduate students are providing care in their clinical rotations are being asked to complete the self-report instrument. Their scores will be compared with the SSOC score assigned by the students in their nursing care plans to determine if the two forms of scoring SSOC are measuring the same construct.

References

Antonovsky, A. (1987). *Unraveling the mystery of health: How people manage stress and stay well.* San Francisco: Jossey-Bass.

Antonovsky, A. (1993a). The implications of salutogenesis: An outsider's view. In A. Turnbull, J. Patterson, S. Behr, D. Murphy, J. Marquis, & M. Blue-Banning (Eds.), *Cognitive coping, families, and disability* (pp. 111-122). Baltimore, MD: Brooks.

Antonovsky, A. (1993b). *The sense of coherence: An historical and future perspective.* Unpublished manuscript.

Antonovsky, A. (1993c). The structure and properties of the sense of coherence scale. *Social Science Medicine, 36,* 725-733.

Antonovsky, A., & Sourani, T. (1988). Family sense of coherence and family adaptation. *Journal of Marriage and the Family, 50,* 79-92.

Artinian, B. (1991). The development of the intersystem model. *Journal of Advanced Nursing, 16,* 194-205.

Brim, O., & Kagan, J. (1980). *Constancy and change in human development.* Cambridge, MA: Harvard University Press.

Budd, K. (1993). Self-coherence: Theoretical considerations of a new concept. *Archives of Psychiatric Nursing, 7,* 361-368.

Burr, W. (1991). Rethinking levels of abstraction in family systems theories. *Family Process, 30,* 435-452.

Esslinger, P. (1994). *A predictor of health status: Situational sense of coherence.* Unpublished manuscript, Azusa Pacific University, Azusa, CA.

Hill, R. (1949). *Families under stress.* New York: Harper & Row.

Kohn, M., & Slomczynski, K. (1990). *Social structure and self-direction: A comparative analysis of the United States and Poland.* Cambridge, MA: Basil Blackwell.

Reiss, D. (1981). *The family's construction of reality.* Cambridge, MA: Harvard University Press.

Rubin, Z. (1981, May). Does personality really change after 20? *Psychology Today,* 239-243.

Shye, S. (Ed.). (1978). *Theory construction and data analysis in the behavioral sciences.* San Francisco: Jossey-Bass.

3

Acquiring a Worldview

The Developmental Environment

Barbara M. Artinian
Darlene E. McCown

Cultural Context of Development
Transformational Processes: The Symbolic Environment
Developmental Outcomes
Clinical Example

A worldview or vision of life is "a framework or set of fundamental beliefs through which we view the world" (Olthuis, 1989, p. 29). It is the "integrative and interpretative framework" for judging life events and "the standard by which reality is managed and pursued" (Olthuis, 1989, p. 29). As such, it has a major impact on the health beliefs and practices held by individuals and communities. It also strongly influences the orientation the person has to life as being comprehensible, meaningful, and manageable—that is, the sense of coherence as defined by Antonovsky (1987). A worldview is socially acquired within a cultural community and is shared by that community. The social organization it provides for behavior within that group is learned by children during the process of socialization. In the Intersystem Model, the *developmental environment* is the term for the arena in which a person's worldview is acquired. The purpose of this chapter is to describe the formative processes by which a person transforms a social environment through his or her particular biological, psychosocial, and spiritual subsystems at each stage of maturational development to become the person who enters the health care delivery system. Using the Intersystem Model (Artinian, 1991), the nurse evaluates the effects of the developmental environment by assessing the client's biological, psychosocial, and spiritual subsystems.

Cultural Context of Development

Leininger (1991) defines culture as the "learned, shared, and transmitted values, beliefs, norms, and life practices of a particular group that guide thinking, decisions, and actions in patterned ways" (p. 147). Because each culture is shaped by particular social and economic realities, each has a different pattern of values, beliefs, norms, and life practices. It follows, therefore, that each cultural group has a different frame of reference for viewing health and health care. In addition, individuals within a culture also view health and health care differently from others within their own cultural group because of their unique characteristics. For example, socioeconomic status allows differential access to health care, and specific biological and social characteristics provide different experiences in social interaction.

Because the response to visible physical characteristics forms the matrix of human interaction, personal appearance greatly influences the feedback the person receives from others and therefore the sense of self of each person. Biological differences exist both between cultural groups and within groups. Just as others respond to a person's biological makeup, the person also responds to self as a person. Mead (1934) describes this as the "I" reflecting on the "me" as the person learns to treat self as an object. For this reason, Kuhn (1974) considers the person's biological system to be his or her internal environment because "the individual learns and makes decisions about his own body by the same processes as for the world outside his skin" (p. 40). Cultural values of what is appropriate appearance are internalized and the person views self through the same lens as the dominant culture, even if that means devaluing of self. Thus, the child with an acceptable appearance grows up in a very different maturational environment from the handicapped or physically less appealing child.

Transformational Processes: The Symbolic Environment

Symbolic Interaction Theory

Kollock and O'Brien (1994) state that the main agenda of symbolic interactionists is to "pursue the question of how the human incorporates existing cultural knowledge into private thought and how this knowledge is acted on, affirmed, and modified in the course of everyday interactions" (p. 32). During development, people learn to comprehend, comment on, and organize behavior in terms of abstract representations of what they understand the rules or patterns to be that operate in a specific situation. These rules are organized through language, which provides categories for the salient aspects of culture. Mead's (1934) contribution to symbolic interactionism is the basic assumption that language is the basis of thought. It is through the process of learning language that "the mind develops and becomes structured in a manner that reflects the individual's culture" (Kollock & O'Brien, 1994, p. 61). It is through reflection on the rules of a culture that individuals can guide their behavior and therefore be socialized into a culture in terms of a shared understanding of the rules.

The reiterative process whereby individuals learn to function within culturally specific patterns is evident in that members of a cultural group attempt to mold the environment to meet their own goals. Their goals are based on socially constructed meanings, however, so that what is judged to be important and how to attain it are shaped by the cultural group within which they interact. Because of this, the major focus of interaction is on the negotiation of meanings in a particular situation, so as to apply culturally assigned meanings and thereby to get others to respond in a desired way. This is possible because of shared symbolic patterns of meaning that allow cues, such as physiological features, types of clothing, or the contextual setting, to be interpreted in a predictable way. When a situation is ambiguous, the person has no abstract representations available within which to interpret cues and, therefore, ambiguity is very stressful.

It is a major tenet of symbolic interactionism that people "act toward things on the basis of the meanings that the things have for them" and that these meanings are socially derived (Blumer, 1969, p. 2). Behavior is based on a subjective interpretation of events rather than on a direct response to the event itself—that is, experience is filtered and modified through the process of symbolic thought. Although the thought process of each person is private, the language patterns through which thought is organized are derived from interaction with others in the culture. Therefore, they are socially shared within a culture. For example, a Western nurse, in interacting with a mother from the same culture who had just been told that her son had leukemia, shared in the grief process but only said, "I know how much a child means to his parents." In another culture, comfort may have been offered by touching or hugging.

The experience and expression of emotion show great variability across cultures. Meltzer and Herman (1990) point out that social and cultural influences produce distinct "vocabularies of emotion" in different societies, and these designate the feelings one can expect to experience in given situations

(p. 186). Shott (1979) has identified the socialization process through which feelings are constructed:

> Within the limits set by social norms and internal stimuli, individuals construct their emotions; and their definitions and interpretations are critical to this often emergent process. . . . It is the actor's definitions and interpretations that give physiological states their emotional significance or nonsignificance. (p. 1323)

By learning the feeling norms or rules of a culture, persons have a frame of reference for naming what they feel and knowing how to respond. They also have a frame of reference for understanding how the other is responding and are able to take the emotional role of the other by putting themselves in the other's position and taking that person's standpoint.

The way an event is stored in memory is based on the way in which the event was interpreted and the meaning assigned to it. This is mediated through the cultural and personal values of the person, which stem from experience. For example, the diagnosis of cancer could be accepted as God's will, as a fitting end to a good life, as a tragic sentence on one who is too young to die, or as a replay of an experience of another family member. It is the process of assigning meanings that determines how a person will respond to a situation.

In addition to interpreting situations in different ways, culture also influences responses to illness. In a grounded theory study of Canadian Indian family management of persistent middle ear disease, Wuest (1991) found that harmonizing was the major strategy used in the process of learning to manage the illness. The native families harmonized by integrating the experience into their reality rather than attempting to direct or take charge of the experience as did the Caucasian subjects.

Interaction in any society also requires the ability to differentiate among the many social roles and to function adequately in a variety of them. To do this, the individual must know the interrelationships between his or her role activities and those of others in counterpositions. Mead (1934) said that this is possible because of an understanding of the generalized set of expectations and attitudes held by a particular culture. These expectations and attitudes provide a working consensus that is a publicly agreed-on definition of the situation that guides interaction in that situation. Individuals learn these guidelines during socialization, and Mead calls this understanding the *generalized other*. This notion of the generalized other enables individuals to know what is expected of them in any social context. By taking the role of the generalized other, persons are able to interpret responses of others from the point of view of societal norms. Therefore, it helps them plan and evaluate their own behavior.

The concept of the generalized other provides the mechanism for ensuring stability in a culture. Change also occurs, however, that is consistent with the larger system so that cultural integrity will not be challenged. Therefore, changes must be circumscribed and emerge from the previous states. Ford and Lerner (1992) state:

> At any moment, the existing states of the system provide facilitating and constraining conditions that define the possibilities for, and the probabilities of, different kinds of potential changes. . . . Each change in system states alters those conditions and thus alters the next set of possible and probable changes. (p. 39)

The facilitating and constraining conditions include factors such as personal characteristics, family characteristics, current state of health, socioeconomic status, value judgments, and personal goals. These unique individual and contextual conditions "promote individual differences in development and foster or diminish the probability of different kinds of developmental path-ways" (Ford & Lerner, 1992, p. 41). Each person has ideas about preferred outcomes and, within the constraints of the particular environment, organizes behaviors to pursue a pathway expected to lead to the desired outcomes.

Socialization Into a Culture: The Development of Faith

The content of culture is not immediately available to the child. During the socialization process, it is progressively learned within his or her maturational level. One aspect of culture that is learned is the cultural and religious value system of a particular culture. In the process of maturation, the person acquires the "preferred patterns of belief and behavior of a particular culture" (Ford & Lerner, 1992, p. 52). As a way of illustrating the interaction of maturational factors and socially available norms, McCown has outlined the development of faith based on the work of Fowler (1981, 1984). This is given as an example to show how any behavioral pattern is acquired and modified by the develop-ing person in a particular culture.

Undifferentiated Faith

The first 2 years of life are critical because the foundation for faith is established. Even prior to birth, during the prenatal period, the health, love, values, expectations, and preparations of the parents for the child establish a basis for the home environment and security available to the new baby. During the first 2 years, a sense of trust in the world and life develops. These years are characterized by a search for conservation or a sense of object permanency in the world. Children build trust in concrete, life-sustaining, and soul-enhancing experiences and in people. These in turn foster trust in the abstract elements of life—God, prayer, and spiritual concepts.

Intuitive-Projective Faith—Stage 1

The first formal stage of faith is *intuitive-projective*. Children ages 2 to 7 years intuit meaning to life. Children formulate imaginary ideas about God, life, and death. They hold an anthropomorphic view of God and describe God with specific physical characteristics (Shelly, 1982). These youngsters hold a punishment/reward concept of morality as defined by Kohlberg (1968). God may be seen as the source of reward or cause of punishment. Children of this age often obey God to avoid punishment.

Young children understand and enjoy simple narrative stories. These stories must be selected carefully, however, to enhance a view of a kind and loving God. They should foster the sense of security, love, and protection afforded by the supreme power in life. The true meaning of religious symbols and acts may be vague to young children, but the ritual aspects of religion provide concrete acts and meaningful consistency.

During these early years of faith development, the child imitates models of faith observed in parents and significant others. Parents often represent Godlike characteristics (omnipotence, omnipresence, etc.) from the child's perspective. Role models of faith are crucial in helping young children establish positive attitudes toward deity and faith in the world of tomorrow.

Mythical-Literal Faith—Stage 2

The faith of school-age children, 7 to 12 years old, is described by Fowler (1981) as *mythical-literal.* For them, the world is literally understood. The ability to think in concrete operations gives rise to a sense of causality and a diminishing of imaginative explanation. This faith stage is characterized by a reliance on the beliefs of authority figures, such as parents, teachers, and clergy.

Just as school-age children insist on rules in a game, these youngsters literally apply religious beliefs. Their sense of moral judgment is based on reciprocal fairness. They desire to please God for the benefits involved (e.g., heaven and health).

Children 7 to 12 years old, however, are also able to take on the perspective of another, even God, and recognize their own capacity for evil and sin. They are able to seek forgiveness and experience conversion. Biblical stories and narrative accounts are a major means for transmitting religious content to school-aged children. They attend carefully to minute details of sequence and character. Again, wise selection of stories is important in developing accurate and positive religious beliefs.

Synthetic-Conventional Faith—Stage 3

Stage 3, *synthetic-conventional faith,* is usually observed in adolescence (12+ years). Adolescence is characterized by the emergence of logical reasoning and comprehension. The teenager begins to think about his or her own beliefs and seeks to understand himself or herself, others, and the world and find meaning and values for life.

Usually an adolescent takes on the faith belief system of his or her community and peer group. This faith stage is "conformist" in that it is tuned to the expectations and beliefs of significant others. (This stage correlates with Kohlberg's [1968] "interpersonal concord" moral reasoning.) The beliefs that are held are deeply felt but usually are not independently or deeply examined.

The youth is able to formulate an ideal self and strive for that self. The ideal self portrayed by religious writings provides a high model for the teenager. Faith offers a basis for ideality and outlook. Religion contains a body of history and narrative accounts that provide a way to understand God.

Individuative-Reflective Faith—Stage 4

Individuative-reflective faith is often observed in young adulthood (20+ years). This stage reflects a greater maturity of faith and movement away from the group conformist faith to a personally owned perspective. The primary task of this stage is to take on personal responsibility for one's own beliefs, commitments, and lifestyle. A meaningful individual outlook is established within a self-selected group.

An important task of this faith stage is to formulate answers to basic religious questions—for example, "Who am I?" The young adult carefully defines and delineates the boundaries and symbols of his or her faith system. Previous acceptance of parental community or peer beliefs is critically reexamined. This is a phase of deliberate selection of a faith style and exclusion of other choices.

Conjunctive Faith—Stage 5

Fowler (1981) calls Stage 5 *conjunctive* faith (35+ years). People at this stage awaken to the realization that truth is more multifaceted and complex than previously recognized. In this stage, frequently a willed submission to the great stories of faith takes place. There is often a reclaiming and reintegration of past faith into present circumstances. Conjunctive faith involves going beyond the carefully delineated Stage 4 (young adult) faith system. People at midlife know the realities of life and personal defeats. Stage 5 faith addresses the uniting of the contradictions in life and the conscious and unconscious aspects of faith.

Conjunctive faith expresses a new openness to the truths of the "other." A sharing of beliefs between people of different faiths makes each vulnerable to the other. The strength of this stage rests in the capacity to acknowledge and accept the powerful spiritual meanings of one's own faith and one's own group as truth while at the same time recognizing their relativity. Stage 5 faith individuals appreciate symbols, myths, and rituals because they understand the depth of reality to which they refer.

Universal Faith—Stage 6

The final stage of faith, seldom realized, is *universal faith.* At this point in faith, the person "lives as though the kingdom of God is here now." Life is both loved and held too loosely. The believer is ready to spend and be spent for the cause and possibility of others receiving meaning and identity in their own lives. Stage 6 persons are "contagious" in creating faith communities. There is a commitment to justice and a selfless passion for a transformed world with divine intentions. As representatives of this stage of faith, Fowler points to Martin Luther King, Jr., Mother Teresa of Calcutta, and Gandhi. Two overriding characteristics of this stage of faith are universal love and inclusiveness. Death is not feared for these individuals; it is incorporated into their living experience. Clearly, not everyone reaches the final stages of faith.

Developmental Outcomes

The personal goals of individuals, their genetic makeup, their emotional and rational characteristics, their ethical or religious inclinations, the interactional processes prevalent within a culture, and the contextual constraints described above work together to form the individual at each stage of development. It is through the process of living that individuals discover who they are and what is important to them—in other words, they develop a worldview. A worldview is formed within a particular cultural tradition and interpersonal context. This worldview gives "reason and impetus for deciding what is true and what really matters in our experience" (Olthuis, 1989, p. 29). Two outcomes of a particular worldview that are especially important in the health care delivery system are the orientation a person has to life as being comprehensible, meaningful, and manageable (sense of coherence as described by Antonovsky, 1987) and the formulation of a health belief system.

As children mature within a particular social community, their educational, work, and family experiences will shape how they view the world. If their life experiences have been consistent, they will view the world as comprehensible; if there has been a balance between overload and underload, they will view the world as manageable; and if there has been the opportunity to participate in socially valued decision making, their sense of the meaningfulness of this world will be strengthened (Antonovsky, 1993, p. 119). This view of the world shapes the appraisal of a particular illness experience and guides the person's response to it.

The health belief model offers an explanation of how health behaviors and attitudes toward health influence the delivery of care. It includes concepts such as *perceived seriousness, perceived susceptibility,* and *cue to action* (Becker, 1974). It incorporates many responses to the health-illness situation, such as how individuals conceptualize physical and mental health or illness; what their beliefs are about ways to prevent illness or promote health; what they think causes illness and their ideas about how to treat illness and their preferences for self-care and home remedies; their use of lay healers or professional health providers; and their ways of describing or expressing illness.

It is through an understanding of the factors influencing development experienced through the biological, psychosocial, and spiritual subsystems of the person, the family, the institution, and the community that nurses are able to understand the worldview of their clients. Such an understanding helps them to provide culturally appropriate care through understanding aspects of cultures such as the meaning of special verbal and nonverbal communication styles, the meaning of space and time to the client, the hierarchal organization of the culture, and an orientation to nature that is fatalistic, harmonious, or seeking to master it (Giger & Davidhizar, 1991). This knowledge enables nurses to mutually plan and implement health care with the client in a culturally perceptive manner. Although this chapter has focused primarily on the influences that have shaped the individual, the same influences shape the social groups within the culture because they are made up of individuals socialized within that culture who experience development over time. By focusing on an understanding of the developmental world of the clients from their perspective and on an understanding of the nurse as a person shaped by his or her culture, an ongoing process of negotiation is possible within the health care delivery situation that is carried out in the situational environment.

Clinical Example

The following analysis of the developmental environment of a geriatric client that made her the person who was ready to plan a visit to her mother is based on an experience recorded by Maura Ryan in her article "Olga's Reason to Live" (1995).[1] The effects of the developmental environment are evaluated by assessing the biological, psychosocial, and spiritual subsystems of the client.

Focus of Situational Interaction

A gerontological nurse practitioner, Maura Ryan, is accompanying her client, Olga, on a visit to Olga's elderly mother, who resides in a nursing home. Olga wonders if she has the courage or strength to go through with the visit.

Developmental Environment

Biological Subsystem

Olga, 77, was run down by a speeding cyclist 9 months earlier and had spent 3 months in an intensive care unit ventilator dependent and suffering from pneumonia and urinary tract infections. She lost weight but gradually recovered enough to be transferred to a rehabilitation facility. While there for 2 months, she was fitted with a built-up shoe for her right leg and she learned to walk using a walker, although she still needs some assistance. She now weighs 75 pounds and appears thin but she has dressed up for the visit and has applied light pink lipstick and blush and has restyled her gray hair. She has a healing midline scar around her neck. She is almost blind.

Psychosocial Subsystem

Olga and her mother, who is 99, fled Nazi Germany. where the rest of their family perished. She lives in a senior apartment complex that is within taxi distance of the nursing home where her mother resides. She and her mother feel that they are survivors and that they are all that the other has. During her period of illness, she was usually unaware of her surroundings, but occasionally she visualized her mother and realized that she could not leave her alone. She said, "I was all that my mother had. How could I desert her after all we'd been through together?" Her native language is German and she uses it to talk with her mother. She is sensitive to the needs of her mother and worried that a phone reunion would be overwhelming to her.

Spiritual Subsystem

Although Olga and her mother are Jewish, Olga gives no indication that faith in God was important to her in the recovery process or in her present situation. More needs to be learned about the role that religion has in her life and the support she receives from the Jewish community.

Note

1. From Ryan, M. (1995). Olga's reason to live. *American Journal of Nursing, 95*(5), 34-35. Adapted with permission from the American Journal of Nursing.

References

Antonovsky, A. (1987). *Unraveling the mystery of health: How people manage stress and stay well.* San Francisco: Jossey-Bass.

Antonovsky, A. (1993). The implications of salutogenesis: An outsider's view. In A. Turnbull, J. Patterson, S. Behr, D. Murphy, J. Marquis, & M. Blue-Banning (Eds.), *Cognitive coping, families, and disability* (pp. 111-122). Baltimore, MD: Brooks.

Artinian, B. (1991). The development of the intersystem model. *Journal of Advanced Nursing, 16,* 194-208.

Becker, M. H. (Ed.). (1974). *The health belief model and personal health behavior.* Thorofare, NJ: Slack.

Blumer, H. (1969). *Symbolic interactionism.* Englewood Cliffs, NJ: Prentice Hall.

Ford, D., & Lerner, R. (1992). *Developmental systems theory: An integrative approach.* Newbury Park, CA: Sage.

Fowler, J. (1981). *Stages of faith.* San Francisco: Harper & Row.

Fowler, J. (1984). *Becoming adult, becoming Christian.* San Francisco: Harper & Row.

Giger, J., & Davidhizar, R. (1991). *Transcultural nursing.* St. Louis: C. V. Mosby Year Book.

Kohlberg, L. (1968). Moral development. In *International Encyclopedia of Social Science.* New York: Macmillan-Free Press.

Kollock, P., & O'Brien, J. (1994). *The production of reality: Essays and readings in social psychology.* Thousand Oaks, CA: Pine Forge Press.

Kuhn, A. (1974). *The logic of social systems.* San Francisco: Jossey-Bass.

Leininger, M. (Ed.). (1991). *Culture care diversity and universality: A theory of nursing.* New York: National League for Nursing Press.

Mead, G. (1934). *Mind, self, and society.* Chicago: University of Chicago Press.

Meltzer, B., & Herman, N. (1990). Epilogue: Human emotion, social structure, and symbolic interactionism. In L. Reynolds (Ed.), *Interactionism: Exposition and critique* (pp. 181-225). New York: General Hall.

Olthuis, J. (1989). On worldviews. In P. Marshall, S. Griffioen, & R. Lieuw (Eds.), *Stained glass, worldviews, and social science.* New York: University Press of America.

Ryan, M. (1995). Olga's reason to live. *American Journal of Nursing, 95*(5), 34-35.

Shelly, J. (1982). *The spiritual needs of children.* Chicago: Inter Varsity Press.

Shott, S. (1979). Emotion and social life: A symbolic interactionist analysis. *American Journal of Sociology, 84,* 1317-1334.

Wuest, J. (1991). Harmonizing: A North American Indian approach to management of middle ear disease with transcultural nursing implications. *Journal of Transcultural Nursing, 3,* 5-14.

4

Context of Involvement

Situational Environment

Judith R. Milligan-Hecox
Margaret England
Barbara M. Artinian

> **Communication of Information**
> **Negotiation of Values**
> **Organization of Behaviors**
> **Discussion**
> **Clinical Example**
> **Conclusion**

It is in the situational environment that nurse and client meet. All the unique characteristics of the nurse and of the client as well as the characteristics of the setting interact to make the experience what it is. Different nurses with differing life experiences would make the encounter a different type of experience for a particular client, as would differing characteristics of the setting. Using the Intersystem Model, each nurse would attempt to understand the problems that the clients are experiencing and work with them within the constraints of a particular setting to bring about a coherent adaptive response. This is done through communication of information, negotiation of values, and organization of behaviors by nurse and client. The goal is to enhance the client's situational sense of coherence (SSOC). This goal is influenced and modified by the quality of intersystem interaction in the situational environment.

The situational environment, as defined in the Intersystem Model, encompasses all the details of the encounter between the nurse and the client. Included in this concept are the place, time, circumstances, and motivational state and receptivity of both nurse and client. Constructs such as perceptual accuracy and uncertainty of the situation influence nurse-client encounters. Situational constraints of the particular environment also influence the repertoire of behaviors that a person uses and limit the outcomes available. The Intersystem Model provides a systematic way for nurse and client to come together to explore the definitions of the present situation so as to negotiate a plan of care for the client that is acceptable to both of them. Achievement of the goals of the plan of care is measured as changes in the client's SSOC.

Use of the Intersystem Model in interaction between client and nurse provides a process to encourage mutual sharing. The nurse and client may meet for the first time as strangers or they may have an established relationship. In either case, communication pathways need to be adequately established. Using the assessment of the biological, psychosocial, and spiritual subsystems to provide a database about the client's knowledge, values, and behaviors, the contextual meaning of the particular situation to the client can be identified. With this information available, mutual goal setting can be initiated. The mutual understanding that occurs during the interaction forms the backdrop for the next situational encounter.

Nurses bear initial and ongoing responsibility for establishing relatedness with clients. The client approaches the interaction because of a need for assistance in matters of health and well-being. The client's mechanisms and strategies to promote coping, however, are important to the healing process. It is important for the nurse to assess how the client is interpreting the illness and what he or she believes is appropriate treatment. Effective nursing requires that nurses become skilled in recognizing and responding to the client's need for help and become sufficiently involved to explore the client's interests.

Elements of nurse-client interaction that may influence changes in the interactional process are, at best, only briefly described in the literature (Kristjanson & Chalmers, 1990). This chapter will explore many of the interactional elements that pertain to enhancement of a client's SSOC.

Communication of Information

Nursing takes place within the context of nurse-client interactions. Basic input into a system of nursing service includes the internal resources the nurse and client bring to the situation and their interpretation of the health situation. This interpretation is often clouded by the situation's ambiguity. The ability to communicate between the nurse and client is important to clarify the situation. Communication within the Intersystem Model begins with a clear identification of the patient's needs, as described by Orlando (1961). Often, patients initially fail to express their needs clearly and the nurse must search for the meaning of the patient's request for help. Orlando states,

> First, the nurse must take the initiative in helping the patient express the specific meaning of his behavior in order to ascertain his distress. Second, she must help the patient explore the distress in order to ascertain the help he requires for his need to be met. (p. 26)

Failure to do this results in the nurse's concentrating on a need or problem the patient may not see as important at the time. Thus, the intervention does not meet the need of the patient. When the nurse is willing to understand the immediate concern of the patient, however, this is interpreted by the patient as caring and leads to further communication.

Effective nursing requires that nurses be able to communicate therapeutically with clients, families, and society. Such conversation occurs through the use of multiple combinations of therapeutic communication techniques such as those described by Hays and Larson (1965). Such techniques are designed to expose clients to situations and messages that will lead ultimately to more concordant experiences in health and healing (Ruesch, 1963).

Every situation consists of unique features. Such features include abstract notions of individual consciousness, environmental shaping, and the relationship between them (Karl, 1992, p. 5). Karl states that a common language is necessary to bridge the gap between inside and outside, or between the individual and the environment. A number of factors influence communication. Among these are nurse characteristics such as the ability to inspire confidence, degree of self-coherence, perceptual accuracy, ability to provide anticipatory guidance, and ability to manage uncertainty.

Confidence of Nurse

The nurse assisting a client requires internal resources to be able to provide the caring and communication skills necessary for the interactive process. Effective nursing requires that nurses be confident. Self-confident, competent nurses are more likely to instill confidence because they can and will look after the needs of their clients. Such nurses can energize and convince clients of their own ability to cope effectively with each development of their situation. Clients become convinced when they learn that they can count on their nurses to respect them and give them accurate information and feedback. They somehow become convinced when they see that their nurses do help them recognize and develop their own resources.

Assurance, trust, and reassurance are three elements of confidence (Kirk, 1992). Assurance refers to a positive declaration, a pledge or guarantee intended to give confidence to others. Trust refers to the expectation that others will behave in one's best interests even when there is increased risk to his or her well-being. Reassurance is an intervention designed to reduce anxiety and uncertainty and to restore confidence (Gregg, 1955; Teasdale, 1989).

Self-Coherence of Nurse

Gaut (1992) has stated that the more the practitioner is rooted in sources that animate the practitioner's being, the greater will be his or her capacity to "be there" for the client. This capacity to be there involves being balanced or self-coherent. To be balanced means that one has to take care of the body, mind, spirit, and relationships. It is important for nurses to know those balances that enhance professional presence and how they can be augmented and shared with others.

According to Budd (1993), self-coherent individuals are aware of their own sources of inner strength. They also are aware of their transactions with the environment and can integrate these transactions with their motivations and goals for living.

Perceptual Accuracy

The nurse needs to be able to interpret what the situation means to the client. This requires perceptual accuracy in understanding the situation. Perceptual accuracy refers to the ability to pick up on and communicate the thoughts, perceptions, feelings, meanings, and actions of others in their situation.

According to Orlando (1972), an organizing principle for nursing is finding out and meeting the client's immediate need for help. Recognition of need for help calls for perceptual accuracy. Somehow nurses must be able to elicit the burdens, sensations, affects, cognitions, and interactions of their clients rather than automatically attributing a string of meanings to the client's behaviors. They must be able to organize, interpret, and transform what they observe about their clients' situation into a meaningful framework.

According to the principle of perceptual accuracy, nurses develop an iterative, self-correcting mode of inquiry with clients. That is, they periodically or repeatedly articulate all or a part of a client's behavior, perceived thoughts, and apparent feelings. Nurses, then, ask the client or their surrogate caregiver(s) or both to verify or correct their description of the client in the context of his or her situation.

A beautiful example of perceptual accuracy was recently reported in the news media. An intensive care unit (ICU) nurse caring for a patient with a head injury became concerned about the aloneness of the unknown victim. The patient was only slightly responsive to verbal stimuli and unable to provide any history. Usual methods to identify the patient had been unsuccessful. Accordingly, the nurse used a unique strategy for communicating with the patient. She began to sing a song that included all of the names of the states in the United States. She noted that the patient showed signs of arousal at the mention of Arizona. She then began to recite names of all of the cities she knew in Arizona. At the mention of Scottsdale, the client again showed signs of arousal. With this information, the institution was able to concentrate its search for the woman's identity in Scottsdale and was able to make contact with her family. This intervention illustrates the ability of one nurse to move far beyond the usual, normative expectations for nursing care so as to communicate with the patient in a truly perceptive manner.

Ability to Provide Anticipatory Guidance

Another communication technique is that of anticipatory guidance. The nurse can enhance the client's decision-making process by carefully choosing the organization, delivery, and documentation of information given to the client based on her expert knowledge of the situation. By providing needed information at the appropriate time, the nurse will facilitate the ability of clients to plan self-management and self-care and improve their ability to carry out everyday activities of self-care.

Ability to Manage Uncertainty

Effective nursing requires that nurses inform clients and act in ways that will increase clients' sense of certainty (Teasdale, 1989). This sense of certainty has the potential to help clients in situations of stress to understand, prepare for, and plan for what they must face in the situation. Such information is reassuring for clients and will help to build their confidence.

Adaptation to an illness experience for people involves managing the uncertainty concerning what will happen. Mishel (1988) defines uncertainty as the inability to determine the meaning of the illness-related events. She further describes it as a cognitive state created when the person cannot adequately structure or categorize an event because of the lack of sufficient cues. Mishel says that uncertainty occurs in a situation in which the decision maker is unable to assign definite value to objects and events or is unable to predict outcomes accurately, or both (Mishel, 1988). Mishel's theory of uncertainty helps to explain the phenomenon of uncertainty and its role in determining the meaning of an illness experience.

Boss (1988) explored the relationship between the perception of a situation and the ability to manage stress. This problem with perception has been described as *boundary ambiguity*. It is especially prevalent in families in which the ambiguity of the illness is long term and severe. "It is the family's perception of the event or situation (the meaning they give it) that is the critical variable in determining the existence and degree of boundary ambiguity" (Boss, 1988, p. 77). If this ambiguity can be lessened, coping strategies can be developed. Even when the outcome of the loss or illness cannot be altered, change can occur within the family coping if the perception of the loss is altered because appropriate categories give order to an otherwise chaotic experience and later some predictable control.

Negotiation of Values

Benner and Wrubel (1989) describe the meaning of an illness as being more than just the immediate situation that the illness brings. The illness is viewed within the whole context of the situation that the illness has precipitated. The effect that the illness will have on the person's expectations, goals, and dreams will have an impact on the response to it (p. 15). Each situation has its own set of meanings to the person involved and cannot be anticipated by another. The personal concerns of the individual will determine the issues of importance in any given situation (p. 88). Within the situation, the person's emotions will begin to be shaped, stress will be experienced, and the possibility for altered coping strategies will emerge.

In addition, the person's understanding of the situation, the meaning attached to it, and the resources available will determine the lines of actions and possibilities that are available to the patient. The nurse must understand these possibilities from the patient's perspective to negotiate values with the patient.

Caring Behaviors

The impetus behind the desire of the nurse to understand the client's perspective and honor client values in planning care is *caring*. As defined by Benner and Wrubel (1989),

> Caring . . . means that persons, events, projects, and things matter to people. . . . "Caring" as a word for being connected and having things matter works well because it fuses thought, feeling, and action. (p. 1)

Without this type of caring, there would be no attempt on the part of the nurse to negotiate values so that care can be organized from the perspective of the patient. It is far easier to provide care in predetermined patterns that one already knows.

Nursing defines the communication process as central to the therapeutic process that constitutes nursing (Rundell, 1991). Nurse and patient come together in a highly interpersonal interactive process when nursing assistance is needed. At the very core and essence of this interaction is the caring that the nurse demonstrates to the patient. A central aspect of caring is that it is always understood in a context (Benner & Wrubel, 1989). Actions that are identified as caring vary with the situation and are central to expert care. For example, it can be "in technical proficiency which requires swift actions or in an expressive action of the patient's uniqueness" (Benner & Wrubel, 1989, p. 5).

Studies of caring behavior in the nurse-patient relationship have identified activities described by clients as being most important to them. These include listening, touching, talking, allowing for expression of feelings, and getting to know them (Winslow, 1994). Other examples of caring from similar studies include accounts of technical proficiency as examples of caring behaviors (Winslow, 1994). Technical proficiency by itself, however, did not constitute caring. In these studies, as nurses provided expert care through technical proficiency, they also explained what was happening to the patient and thus provided reassurance.

A subset of the caring context is the use of empathy, a well-acknowledged important component of the nurse-client interaction. It has been defined as a "process wherein an individual is able to see beyond outward behavior and sense accurately another's inner experience at a given point in time" (Gilje, 1992, p. 62).

Caring requires a unique energy that originates from a source beyond self (Montgomery, 1992, p. 39). This resource is similar to that of spiritual transcendence. The presence of this resource makes it possible for one to sustain the pain and loss that result from forming attachments and undergoing losseswith people during tragic circumstances (Montgomery, 1992, p. 45). Nurses who are able to grow and mature in the philosophical wisdom gained from this transcendence view life with a deep acceptance of what to others may seem senseless and meaningless tragedy (Montgomery, 1992, p. 47).

Professional Values

One problem nurses encounter as they attempt to provide nursing in a caring environment is the idea of professional socialization that warns against getting "too involved with clients" (Montgomery, 1992, p. 39). This seeks to devalue the concept of caring in nursing. At a deep interpersonal level, however, nurses continue to care and get involved with clients with an intensity of depth that seeks to foster care. This does not mean that the nurse is to succumb to the destructive forces of overinvolvement but rather to care in a way that enhances and reinforces professional satisfaction and commitment (Montgomery, 1992). The intersystem relationship that respects the boundaries and supra- and subsystem supports of both the patient and the nurse provides the mechanisms to prevent overinvolvement. Because the nurse does not completely become a part of the client system and understands the limited nature of the interdependence between nurse and patient, he or she is able to maintain an autonomous perspective.

Interpretation of the Situation

Also important to negotiations between the nurse and the client is the internal processing that each brings to the situation. An important consideration for any negotiation between two individuals is to realize that each will interpret the situation based on his or her own understanding. Several social theories have been developed to help explain the social processes present in human interaction: symbolic interaction and frame analysis.

Symbolic Interaction

People are symbol-using creatures who interpret their world according to the meaning they have assigned to it. The term *symbolic interaction* was first used by Blumer (1969) to define an approach to the study of the interaction between two or more people. His interest was in studying the interpretation of situations and what occurred between the stimulus and response of any given situation. This interpretation is based on both the situation itself and the cues that a person derives from it. Using his construct of symbolic interaction, one looks at the meaning that is assigned to a situation and what is the consequent response to it. The following three implications for human response have been identified:

1. People act on the basis of the meaning a situation has for them.
2. Meaning is not absolute; it is socially derived.
3. Social symbols are interpreted through the person's thought process (Kollock & O'Brien, 1994, p. 54).

One method to use in interpreting the actions of other is role taking. Reynolds (1990) states that role taking is the basic sociological process of social interaction. It involves standing outside of one's usual role and placing oneself in another's place. The role taker adjusts his or her behavior in the course of the interaction through the use of significant symbols such as gestures and cues, which are then interpreted to discover what these roles are. According to Reynolds, people role take to make sense out of others' actions and, thus, facilitate their own responses to others in the context of the social situation (p. 207).

Frame Analysis

Another sociological concept important to understanding human interaction is that of *frame analysis* (Goffman, 1974). Frame analysis provides a definition for a situation to participants based on common cultural knowns. All the complexities of history, society, culture, and myth help shape one's understanding of a situation. Each aids in providing personal meaning to the situation.

Interactions between groups are understood by members of the group through an understanding of shared culture. The meaning of the situation is organized, characterized, and identified by each of the participants in terms of shared definitions. Using the concept of frame, a definition is established for one situation; this definition is then used to interpret and describe further situations. When one says that a frame is clear, it means more than that each participant has a workably correct view of the situation. It implies a tolerably correct view of the other's views that emphasizes one's perspective of the other's perspective (Goffman, 1974, p. 338). Frame analysis organizes the involvement, expectations, and limits that a person places on a situation (p. 345).

Inequality of Power

An important factor in the negotiation of values carried out by the nurse and patient is the inequality of power between them. Knickrehm (1994) is developing a theory of inequality in social interactions within the perspective of symbolic interactionism. Because symbolic interaction is "an on-going process involving actors who are communicating their interpretations of the meanings of whatever is being communicated, and their definitions of the situation of the interaction" (Blumer, 1969, p. 66), it is apparent that each actor will have a different definition. Actions taken by the interactants are based on their interpretation of their own definition of the situation and on their interpretation of the perception of the other party's definition of the situation. Knickrehm (1994) refers to the synthesis of the definitions brought to and emerging from an interaction as the *staging* of the interaction process.

One factor affecting the staging is the *awareness context* of the participants, which Knickrehm says is derived from their identities that are developed prior to the interaction (i.e., in the developmental environment). These are "structural properties of social interaction which circumscribe the range of possibilities for human behavior available in any interaction . . . and shape the interaction" (Knickrehm, 1994, p. 11).

The outcome of the interaction process is a mutual definition of the situation that Knickrehm (1994) calls a "negotiated" awareness context. Because each person has a unique self, no two persons have the same awareness context. "Their contexts will differ on the basis of their prior experiences, as these experiences have come to shape their selves" (p. 13). Through manipulation of the awareness context, a mutual definition emerges. Because of the differing ability of each to manipulate the resultant definition, however, one person may have more input into the mutual definition than the other. In other words, one person may be more powerful than the other if power is defined as the ability to do something. Therefore, Knickrehm states that "power is rooted in the differential abilities of actors to manipulate the mutual definition of an interaction situation" (p. 14). Power must be considered as a dimension of interaction. Knickrehm further states,

A person's position in a status hierarchy has an effect on that person's interactions, in that status as perceived by others is used by others in their definition of the interaction situation, thus serving to mobilize the ways in which they respond to the person. Thus power, and so inequality, is not only derived from the process of the formation of self, it is an interactive part of this formation. Inequality, that state in society in which one status is relatively lower than another status, is indeed a quality inherent to social life. (p. 16)

This is an important consideration in attempting to develop a mutual plan of care in the Intersystem Model. The nurse, because of a greater knowledge of the medical aspects of care and institutional policies as well as a more dominant position within the caregiving environment, will have a greater ability to steer the mutual definition of the situation. Therefore, great care must be exercised to ensure that the client's knowledge and values have important input into the negotiated awareness context.

Organization of Behaviors

To understand how it is possible for nurse and client to develop a mutual definition of the situation so that a mutual plan of action can be undertaken, the broad factors that influence the organization of nursing care must be

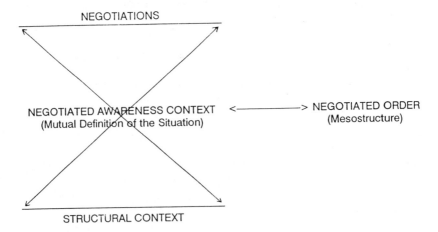

Figure 4.1. Negotiated Order

considered. It is no longer sufficient today merely to examine nurses' attitudes without examining the institutional structure and environmental climate of their surroundings. Although attitudes depend to some extent on the personality of the individual nurse, they also relate to the institutional ecology of the particular setting in which care is administered.

Negotiated Order

Three factors affect the outcomes of communication or negotiation between individuals: (a) structure (the larger societal level within which negotiations take place); (b) process (the ways of getting things accomplished); and (c) the negotiation context (the "relevant features of the setting which directly enter into negotiations and affect their course" [Strauss, 1978, p. 101]).

The structural context includes the stable features of an organization that are the background through which people interact, such as organizational size, type of leadership, rules and policies, organizational goals, degree of professionalization of staff, and focus of resource allocation. Process includes the types of interactions used by persons in the setting, such as bargaining, compromising, fighting, and colluding. The actual negotiation context includes those aspects of the situation that directly affect the negotiations, such as the extent to which information is shared, personal goals, hidden agendas, limited budget, amount of time available, the participants' emotional state and relative power position, and the immediate situational context. Each of these factors will influence the outcome of the negotiations and produce what Strauss (1978) has called *the negotiated order.* The societal expectations for the particular structural context influence how each of the persons in the setting will react. The relationships between the individuals in negotiation and the structural organization have been diagrammed by Artinian and are depicted in Figure 4.1.

Maines (1979) has extended the concept of negotiated order to show how negotiation contributes to the formation of social order and how social order influences the negotiation process. He used the term *mesostructure* to describe the domain in which the dialectical activity takes place between subject and object (negotiation and social order) to form the domain of subject-object unity. He says that the mesostructure (or negotiated order) is the domain in which the negotiation process takes place that mediates between the organizational structure and the interacting persons. It is the domain between the interactional and the structural aspects of the situation.

Effect of the Work Setting

Because the workplace is the context of the negotiated order, it can be seen that the work setting in which nurses practice can possess great power to enhance or inhibit practitioner function. The turmoil of the dysfunctional work arena can inhibit the effectiveness of the person and drain morale (Karl, 1992, p. 7). Pressures of management style, technology, bureaucracy, and finance can blur and devalue the importance of caring practice (p. 7).

The organizational culture of the clinical setting may not encourage nurses to invite patients to talk with them about their problems. A study done by Burnard and Morrison (1988) rated nurses' interaction as more authoritative than facilitative. They found that problems of time, work, and perceived priority constraints were factors that influenced the nurse's perception of type of interactive style.

Nurses who communicate effectively are more able to move beyond the constraints of the institutional environment and help patients to regain health. With the advent of managed care and the denial or postponement of certain services by authorizers, these skills become even more important.

The support of nursing management is an important contributor to effective nurse-patient interaction. It is the nurse manager who is in control of issues such as overwhelming workloads that prohibit time for meaningful nurse-patient interactions. Many nurses identify with the feeling of not being cared about in the work setting as they struggle with a heavy workload and little time to deliver care. A nurse manager who uses strategies such as being sensitive and open to staff, offering help and understanding, and encouraging camaraderie and respect can enhance the caring environment nurses desperately need (Dietrich, 1992).

The setting in which nurse-patient interaction takes place, coupled with the element of who has the dominant role or control, is critical to the flow of information exchange and negotiation in nurse-patient interaction. With today's shift in health care settings, it is vitally important to examine the effect of these settings on nurse-patient interactions.

Specific Health Care Settings

The environment of a specific health care setting has considerable effect on the nurse-patient interaction. Because health care today has shifted its focus from acute care to outpatient or community-based care, nurses need to learn to function in environments in which the nurse-patient interaction has very different dynamics. In this section, differences found in a variety of clinical settings will be examined.

Home care is growing at a phenomenal pace (Stulginsky, 1993). This has brought about significant changes in the situational environment of the nurse's workplace. Nurses who have previously worked in the powerful and invulnerable technological environment of the ICU may now find themselves working in the patient's home (Bowers, 1992). The shift of power is immediately felt.

Nurses in home settings describe the patient as the control center from which the health care providers take direction (Stulginsky, 1993). According to Stulginsky, control belongs to the patient and, therefore, the nurse's power base is lessened. Families feel free to question the nurse's suggestions, and often their care for the patient is based on their own best judgment. This loss of power is often the greatest adjustment that the nurse must make when moving into home care from the institutional setting.

Families and patients see nurses in the ICU environment as having a disproportionate amount of power. This power differential is reinforced by differences in the patient's vulnerability and health status and in the nurse's knowledge and familiarity with the context. In a study by Cooper (1993), the ICU environmental setting was shown to promote power and knowledge differentials between nurse and patient. This inequality was reinforced by differences in meaning that this environmental setting held for the nurse and patient. Because ICU patients were in a vulnerable health state, they were unable to exact any personal control over their situation.

Another environmental problem in the ICU setting is the conflict between technology and caring. Studies of the paradoxical nature of technology and care in the ICU environment have described it as a feeling outside of the ordinary world of human experience. Cooper (1993) says this is because health care technology is designed to overcome those human experiences that define ordinary existence, including vulnerability and uncertainty, and because nurses sometimes identify more with powerful, invulnerable technology rather than with the powerless, vulnerable patient (Cooper, 1993). This shift reflects a natural tendency to take on the dominant characteristics of the ICU environment in which the nurse lives and works.

Oncology nurses have described their environmental setting as "being on the front lines of a war against death, disfigurement, and intense suffering" (Cohen & Sarter, 1992, p. 1482). This setting creates tremendous stress as nurses struggle with multitudinous physical needs of the patients as well as a deep concern for patients and families. They describe their environment as that of situations of unexpected crisis dealing with patient suffering and death (Cohen & Sarter, 1992). Nurses in this setting feel that they must be constantly vigilant to manage frequent—and sometimes cycles of—unexpected physically and psychologically based crises.

The environmental setting of the security forensic hospital is another area that affects the extent of power and control in nurse-patient interactions. Nurses hold keys to locked wards and must enforce control and adherence to rules and policies. These same nurses, however, attempt to maintain the therapeutic caring environment in which healing can occur (Caplan, 1993). Forensic patients have unique characteristics that influence the treatment environment and can pose real physical danger for nurses. Caplan, however, reports that nurses perceived that their important role was one of planning activities, schedules, and routines for the patients. On the other hand, the patient's perception of the nurse's role was quite different. Patients rated the amount of control maintained by the nurse as above average. Clearly, the patient's perception of the situational constraints in this setting differed from the nurse's perceptions.

Discussion

When the Intersystem Model is used to direct nursing practice, the nurse and patient together clarify the understandings of both and negotiate mutually agreed-on goals. The detector part of the system provides information about the state of a system's environment. It recognizes the powerful forces that play on nurse and patient and thus allows an examination of the interdependent dynamics of interaction both within and between the systems. As the nurse and patient come together in a highly symbolic, meaningful interaction, values are negotiated and behaviors are organized to develop a joint plan of care. This process leads to increasing the situational sense of coherence of the patient and thus enhancing the patient's health.

The patient may experience a stressful health event as one of uncertainty or danger or even deny its existence. The cognitive state of the patient may disrupt his or her capacity to devote the energy needed to adequately understand and manage the situation. When patients experience uncertainty and unpredictable outcomes related to their illness experience, their ability to make plans in the course of everyday living may be hindered. As the patient and the nurse come together to appraise the meaning of the experience, the need for nursing intervention becomes evident.

Often, patients are forced to become intimate with nurses because of their need for counseling, personal care, and test preparation. This intimacy may be different from their pattern of usual social contacts. They also can feel vulnerable because of their dependency on the nurse for physical and emotional care and psychoeducation (Koller, 1988). The therapeutic communication skills of the nurse at this time will be vital to allaying the patient's anxiety.

Using self in a therapeutic manner, the nurse can assist the patient in any situation to increase his or her SSOC. Reassurance is enhanced when people sense that others take them seriously and legitimate their expertise and experiences as being credible. Patients who can trust their nurses are more likely to gain confidence in their own self-worth and abilities for coping and healing. Patients who trust are more likely to develop and use their supports effectively (Grace & Schill, 1986).

The presence of trust enables the patient to attempt to regain and maintain an optimal level of health in the context of the life-threatening illness. For the critically ill patient, this sense of trust may be all that lies between life and death.

Much of nursing calls for the installation of confidence in patients in the face of threat. Patients commonly are uncertain and at a loss as to how to cope in their health situation. A building of confidence or trust between the nurse and patient is important in decreasing the patient's sense of incompetence, which is necessary for successful adaptation to the illness situation. As Reynolds (1990) has so eloquently stated,

> Just as human relationships (the conceptual unit of social structure) and human interaction (the conceptual unit of social process) are but two sides of the same coin, so are their respective matrices (social structure and social process) themselves similarly linked. (p. 223)

It is from this type of systemic thinking that the Intersystem Model describes the interaction of nurse and patient as they learn and plan in complex situations and environments of continuous change.

Clinical Example

It is in the situational environment that the nurse and patient come together to clarify their understanding of the patient's illness situation and to negotiate mutually agreed-on goals. All the characteristics of the patient and of the nurse as well as the structural context of the setting influence the interaction that takes place. Although the Intersystem Model provides a unique framework for assisting the nurse to understand the perspective of the patient and develop a collaborative plan of care, the underlying principles of the model are practiced by many perceptive nurses who truly want to include the patient in the planning of care. An example of this perceptive care is given by Debra Gordon and Sandra Ward in their article, "Correcting Patient Misconceptions About Pain" (1995).[1] The interactions of Gordon and Ward as they work with their patient, Joan Staley, to set goals for the management of her pain are analyzed within the framework of the Intersystem Model.

Focus of Situational Interaction

The nurse asks the patient a question that she has not thought about: "On the 0-to-10 scale, what's your goal? What level of pain relief do you want to achieve?" The interaction that takes place focuses on helping the patient understand what she could expect in terms of pain control and what goals are possible for her and on helping the nurse understand the patient's perspective on pain control. The purpose is to provide a target and sense of direction for the pain management plan.

Assessment

Developmental Environment

Biological Subsystem

Joan Staley, age 42, had a hysterectomy for uterine cancer 48 hours prior to this interaction. She is a patient on a surgical gynecology unit and has been asked frequently to describe the intensity and type of pain she is feeling.

Psychosocial Subsystem

She prides herself on her ability to "tough things out" and has rarely taken analgesics. She has teenage children and has worked hard to keep them off drugs. She believes that the misuse of drugs will cause her to get hooked on them and may lead to their ineffectiveness at a later time.

Spiritual Subsystem

Nothing is known about her spiritual orientation at this time. With further interaction, particularly if her cancer is not cured, she may discuss this with the nurse.

Situational Environment

Intrasystem Analysis of the Patient

Detector Function. The patient believes that some pain is associated with cancer and surgery and is not aware of the negative consequences of pain on recovery. She is not aware of breathing and splinting techniques to relieve pain. She has misconceptions about addiction and pain relief. She believes that the nurses are doing everything they can to control her pain.

Selector Function. The patient would like to be free of pain but does not expect to be. She does not see herself as a complainer. She is more concerned about her future need for effective analgesics than for the pain she is currently experiencing.

Effector Function. The patient is using a patient-controlled analgesia (PCA) pump to give her own morphine but is reporting a pain level of 4 to 7 depending on her activity. Although she is experiencing pain, she is able to do coughing and deep breathing exercises.

Intrasystem Analysis of the Nurse Case Manager

Detector Function. The nurse understands the principles of pain control management and their benefit for comfort and recovery and adverse effects of analgesic drugs. She understands that although the patient expressed satisfaction with the pain relief provided, it may be a result of not providing the patient with enough information to make a proper evaluation. She knows that she could be doing a better job. She understands the need for patients to define a goal for pain relief.

Selector Function. The nurse values knowing patient expectations and preferences for pain control. She values patient teaching so that the patient can be a participant in planning care by knowing how to develop a goal that would assist her to make informed decisions about her care.

Effector Function. The nurse assessed patient's physiologic and behavioral responses to pain by monitoring pain level every 2 to 4 hours and noting drug usage and possible adverse effects of drugs. The behavior of the nurse displays accurate perception, confidence, self-coherence, and the ability to provide anticipatory guidance.

Analysis

Identification of Stressors

Diagnosis of cancer
Pain from recent surgery

Coping Resources

Previous experience with "toughing things out"
Teenagers who are important to her
Available analgesics
Supportive nursing staff

Scoring on Situational Sense of Coherence

Comprehensibility = 1. Patient does not know the benefits of pain relief and the negative consequences of pain. She is concerned about getting hooked and fears she will develop tolerance to medications so they will be ineffective in the future if her cancer is not cured.

Meaningfulness = 1. The nurse believes that a goal for pain management is important, but the patient does not see the importance of having a goal.

Manageability = 2. Since surgery, pain has not been kept within acceptable limits because of inadequate goals. The patient, however, is well educated and able to process the information effectively and can learn to develop adequate goals if she sees the importance of pain relief. The institution has the needed resources for pain management in terms of effective nursing staff, adequate equipment, and analgesics.

Nursing Diagnosis

Inadequate management of pain related to knowledge deficit about adequate pain relief.

Nursing Goals

1. To convince patient that pain relief is beneficial
2. To set up a plan for pain management

Intersystem Interaction

Communication of Information

In response to the patient's misconceptions about the use of analgesics, information was given in a way that would not make her feel foolish. Authoritative literature was provided that would dispel her fears about addiction, tolerance, and side effects. Pillow-splinting and breathing techniques to lessen pain while coughing and deep breathing were demonstrated. This information was accepted because the patient recognized its validity and perceived that the nurse was accurate in her assessment of her need for pain relief. The patient recognized that although this was a new and uncertain situation for her, the nurse had experienced it before and was competent to provide guidance for her.

Negotiation of Values

Although the nurse and patient were far apart in their perception of the value of pain management at the beginning of the interaction, through the caring behaviors of the nurse her value of needing to set a goal for pain relief was recognized by the patient as being appropriate for her and was accepted as her own. When she realized that her valued behavior of toughing it out was actually detrimental to her recovery process, she reinterpreted the situation so that both patient and nurse shared the same interpretation of the value of goal setting for pain management. At the beginning of the interaction, the patient was in a position of low power because of inadequate knowledge. Through appropriate teaching, the patient was brought to the place where she could make an informed decision about her care. She was no longer satisfied with a pain rating of 4 or more. This made it possible for a goal to be determined that would be both reasonable and realistic and acceptable to both patient and nurse.

Organization of Behaviors: Mutual Plan of Care

1. Patient will study information provided by nurse and test the effectiveness of new pain control techniques.

2. Patient will monitor her pain during times of rest and activity to identify a reasonable pain goal linked to activities necessary for her recovery.
3. Patient will evaluate the new program.

Because the patient did not like to focus on pain, with the consent of the surgeon a continuous hourly basal infusion of analgesic was added to her PCA regimen while still allowing her to have personal control over additional medication during activity. The nurse continued to monitor her use of analgesics and asked, "Do you feel we need to adjust your dose?" With reassessments of pain control during the next 12 hours, the patient's self-report of pain was brought down to her desired goal of 2 or 3. She was able to be more active and independent, and by the next day she was able to switch to an oral analgesic. She continued to direct her pain management program and recommended that her dosing schedule be changed based on her activities so that she took the analgesic 1 hour before bathing, ambulating, and family visits. The negotiations that took place within the structural context of the surgical gynecology unit led to a mutual definition of the situation and a negotiated order that was exemplified in the plan for meeting the patient's goal for pain relief.

Evaluation of the Plan: Rescoring on SSOC

Comprehensibility = 3. The patient has full understanding of the pain management program and was able to set a realistic goal for herself and meet it.

Meaningfulness = 3. The patient recognized the value of setting a goal for pain relief in terms of her overall recovery and actively developed her own program.

Manageability = 3. The patient had the resources necessary to manage her pain and the needed assistance to use them. Her pain was maintained at the desired goal.

The management of postoperative pain is just one aspect of providing care for the cancer patient. With the focus on helping the patient accept the need for setting a goal for pain management, many other issues were not addressed. The trust and confidence in the nurse and the institutional setting that were developed during this interactional sequence, however, will become part of the developmental environment of the patient and will influence future interactions. Issues such as further treatment for the cancer, the ambiguity of whether the cancer will become worse, and so on will need attention as they are introduced by the patient. This analysis of the interaction around the need for goal setting for pain relief illustrates that in using the Intersystem Model format it is permissible, based on her intrasystem assessment of the patient, for the nurse to introduce her concerns for the patient that the patient did not even know should be concerns.

Conclusion

The major goal of nursing when using the Intersystem Model is to assist the client in increasing SSOC. Clients with a high SSOC are more likely to heal and stay healthy because they have developed competencies for managing their health situation. That is, they are able to identify and use options to influence the outcome of a situation. They can evaluate their responses in the context of what is normative and what is normative for them and can enact decisions based on what they see as correct information. They are able to endure difficult times and comfort themselves. Finally, individuals who are high in SSOC are able to communicate effectively and influence how others respond to them. Nurses, through their use of self, can fulfill their responsibility by helping individuals recognize, transform, and transcend their human experiences in the situational environment.

Note

1. From Gordon, D., and Ward, S. (1995). Correcting patient misconceptions about pain. *American Journal of Nursing, 95,* 43-45. Adapted with permission from the *American Journal of Nursing.*

References

Benner, P., & Wrubel, J. (1989). *The primacy of caring.* Menlo Park, CA: Addison-Wesley.
Blumer, H. (1969). *Symbolic interactionism.* Englewood Cliffs, NJ: Prentice Hall.
Boss, P. (1988). *Family stress management.* Newbury Park, CA: Sage.
Bowers, I. (1992). Ethomethodology II: A study of the community psychiatric nurse in the patient's home. *International Journal of Nursing Studies, 29,* 69-79.
Budd, K. W. (1993). Self-coherence: Theoretical considerations of a new concept. *Archives of Psychiatric Nursing, 7,* 361-368.

Burnard, P., & Morrison, P. (1988). Nurses' perceptions of their interpersonal skills: A descriptive study using Six Category Interventional Analysis. *Nurse Education Today, 8,* 266-272.

Caplan, C. (1993). Nursing staff and patient perceptions of the ward atmosphere in a maximum security forensic hospital. *Archives of Psychiatric Nursing, 7,* 23-29.

Cohen, M. Z., & Sarter, B. (1992). Love and work: Oncology nurses' view of the meaning of their work. *Oncology Nursing Forum, 19,* 1481-1486.

Cooper, M. C. (1993). The intersection of technology and care in the ICU. *Advances in Nursing Science, 15,* 23-32.

Dietrich, L. (1992). The caring nursing environment. In D. Gaut (Ed.), *The presence of caring in nursing* (pp. 69-88). New York: National League for Nursing Press.

Gaut, D. (1992). *The presence of caring in nursing.* New York: National League for Nursing Press.

Gilje, F. (1992). Being there: An analysis of the concept of presence. In D. Gaut (Ed.), *The presence of caring in nursing* (pp. 53-68). New York: National League for Nursing Press.

Goffman, E. (1974). *Frame analysis.* New York: Harper & Row.

Gordon, D., & Ward, S. (1995). Correcting patient misconceptions about pain. *American Journal of Nursing, 95,* 43-45.

Grace, G. D., & Schill, T. (1986). Social support and coping style differences in subjects high and low in interpersonal trust. *Psychological Reports, 59,* 584-586.

Gregg, D. (1955). Reassurance. *American Journal of Nursing, 55,* 171-174.

Hays, J. S., & Larson, K. H. (1965). *Interacting with patients.* New York: Macmillan.

Karl, J. C. (1992). Being there: Who do you bring to practice? In D. Gaut (Ed.), *The presence of caring in nursing* (pp. 1-13). New York: National League for Nursing Press.

Kirk, K. (1992). Confidence as a factor in chronic illness care. *Journal of Advanced Nursing, 17,* 1238-1242.

Knickrehm, B. (1994, August). *The origins of inequality: A theoretical synthesis.* Paper presented at the 1994 American Sociological Association Meeting, Los Angeles.

Koller, M. (1988). Risk as a determinant of trust. *Basic and Applied Social Psychology, 9,* 265-276.

Kollock, P., & O'Brien, J. (1994). *The production of reality.* Thousand Oaks, CA: Pine Forge Press.

Kristjanson, L., & Chalmers, K. (1990). Nurse-client interactions in community-based practice: Creating common ground. *Public Health Nursing, 7,* 215-223.

Maines, D. (1979). Mesostructure and social process. *Contemporary Sociology, 8,* 542-547.

Mishel, M. H. (1988). Uncertainty in illness. *Image, 20,* 225-232.

Montgomery, C. L. (1992). The spiritual connection: Nurses' perceptions of the experience of caring. In D. Gaut (Ed.), *The presence of caring in nursing* (pp. 39-52). New York: National League for Nursing Press.

Orlando, I. J. (1961). *The dynamic nurse-patient relationship: Function, process and principles.* New York: G. P. Putnam.

Orlando, I. J. (1972). *The discipline and teaching of nursing process.* New York: G. P. Putnam.

Reynolds, L. (1990). *Interactionism: Exposition and critique.* Dix Hills, NY: General Hall.

Ruesch, J. (1963). Synopsis of the theory of human communication. *Psychiatry, 16,* 215.

Rundell, S. (1991). A study of nurse-patient interaction in a high dependency unit. *Advances in Nursing Science, 10,* 171-178.

Strauss, A. (1978). *Negotiations.* San Francisco: Jossey-Bass.

Stulginsky, M. M. (1993). Nurses' home health experience. *Nursing and Health Care, 14,* 476-485.

Teasdale, K. (1989). The concept of reassurance in nursing. *Journal of Advanced Nursing, 14,* 444-450.

Winslow, E. (1994). How patients define caring may surprise you. *American Journal of Nursing, 94,* 57-58.

5

Theoretical Foundations
for the Biological Subsystem

Margaret M. Conger

Biological Subsystems
Clinical Example

The human body is a unity of interrelated biological subsystems that corporately maintain all the physiological functions needed for life. A systems approach to understanding how each subsystem is controlled and how it interacts with both the internal and external environment is the focus of this chapter.

Bertalanffy (1968) described a system as being that which provides an overall structure so that parts that are not directly connected to each other can be integrated into an organized whole. Each subsystem is open to input from other subsystems and provides an output that will affect other subsystems. The activities occurring within the subsystem are referred to as throughput. Throughout this discussion, the inputs, throughput, and outputs of each subsystem will be addressed.

The Intersystem Model uses the biological subsystems as a guide for gathering information for the detector function. Nurses skilled in the use of the Intersystem Model are able to incorporate this information into their client assessments and make clinical decisions based on the information. Understanding the major concepts important to the function of each of the subsystems will enable the nurse to understand the pathophysiological changes in the client's responses. Decisions about the interventions needed to reverse these changes arise from this knowledge. In this chapter, the major concepts important to the functioning of each of the biological subsystems will be reviewed with an emphasis on identifying the theoretical foundations of each.

In the Intersystem Model, the body's subsystems are divided into the following:

1. Regulatory subsystem
   a. Neurological
   b. Endocrine
   c. Immune
2. Circulatory and respiratory subsystem
3. Musculoskeletal subsystem
4. Gastrointestinal subsystem
5. Renal subsystem
6. Reproductive subsystem

Each subsystem itself functions under a series of control mechanisms, or inputs, that directs its function. These inputs can arise from a smaller unit, the cell, or from other subsystems that generate internal regulators. It is through this regulation that a constant internal environment is maintained despite moment-by-moment changes. This maintenance of a constant internal environment despite constant change is known as homeostasis. When a control mechanism ceases to function, or has diminished function, homeostasis is disrupted and a pathological event can occur.

In addition to inputs generated internally by the body, many of the subsystems are also open to inputs from the external environment surrounding a person. Contact with the external environment is maintained through the body's senses such as sight; hearing; smell; somatosensory including touch, heat, and pain; and locomotion. Examples of subsystems that are in contact with the external environment include the gastrointestinal, renal, and respiratory systems. The skin as a covering of the musculoskeletal subsystem also serves to provide communication with the external environment. All but the renal subsystem also have the ability to take in materials from the outside environment. In addition, all these subsystems have functions that can rid the body of products that are no longer needed and that, if retained in the internal environment, could cause harm.

Biological Subsystems

Regulatory Subsystem

The regulatory subsystem consists of three interconnected parts. The first is the neurological subsystem, which responds rapidly to changes in both the internal and external environments. The endocrine subsystem has longer term regulation of many body functions. It is an important adaptive control system to environmental stressors. Finally, the immune subsystem provides protection from foreign materials that invade the body.

Neurological Subsystem

The nervous system is the portion of the regulatory subsystem that responds to minute-by-minute changes in either the external or internal environment. It initiates responses that allow the person to interact with the environment and also serves to protect the person from harm.

Central Nervous System. The central portion of the nervous system consists of the brain and the nerve tracts that carry messages into and out of the subsystem and function in a cognator or conscious thought process. The somatic portion of the nervous system regulates actions that are under conscious control. Stimuli from the external environment are transmitted via nerves to the brain and are processed in the cerebral cortex. There, a stimulus to initiate an action can be developed and transmitted back to the musculoskeletal subsystem for an action to occur.

An example of the nervous subsystem's response to a stimulus can be illustrated by the events that occur when a person touches a hot object. Pain receptors in the skin will be stimulated by the hot temperature and initiate transmission of a nerve impulse to the thalamus. This impulse serves as the input signal into the system. The output signal is a nerve impulse that stimulates the muscles in the area of the body that touched the hot object to contract. The response is to move the body away from the object. This process is known as a withdrawal reflex (Asratian, 1965).

Even while this is happening, the sensory impulse will continue on to the cerebral cortex where the information about the hot object becomes part of conscious thought. Responses can also be initiated by areas in the lower brain stem without conscious involvement.

The nerves involved in this transmission are under a carefully regulated control system that allows a person to live in a world producing multitudinous inputs. The control system for regulating which of these stimuli reach a person's conscious level prevents neurological overload.

Transmission of an impulse from one neuron to the next is through a chemical neurotransmitter, acetylcholine. Other neurotransmitters include serotonin, dopamine, endorphin, noradrenaline, and γ-aminobutyric acid.

The gate control theory of pain management developed by Melzack and Wall (1965) and Casey and Melzack (1967) illustrates how effectively the nervous system can control the reception of impulses from the external environment. Impulses from two types of fibers, thick fibers and thin fibers, are transmitted to an area lying just outside the spinal cord called the substantia gelatinosa. This area serves as a regulatory center. The fate of excitatory impulses depends on what other stimuli are also received at this center. When inhibitory impulses transmitted by the thick fibers predominate, excitatory impulses transmitted by the thin fibers will be blocked and the pain impulse will not be transmitted to the cerebral cortex. No sensation of pain will be perceived by the person.

Regulation of pain impulses can also be affected by neurotransmitters such as endorphins released within the brain. Within the brain, both inhibitory postsynaptic potential (IPSP) and excitatory postsynaptic potential (EPSP) neurotransmitters are released (Meinhart & Mccaffery, 1983). Stimulation of IPSP causes release of endorphins that in turn block the transmission of pain excitatory impulses in the substantia gelatinosa. On the other hand, stimulation of the EPSP neurons will result in the pain sensation being felt.

Autonomic Nervous System. The neurological subsystem also includes the autonomic system, which responds to inputs from the internal environment and controls many of the body's other internal subsystems without conscious control. The autonomic subsystem consists of the sympathetic and parasympathetic subdivisions. Generally, the sympathetic subsystem functions under duress to protect the individual by preparing the person for "flight or fight." For example, if the external senses, such as sight or hearing, alert the person to a possible danger in the external environment, the sympathetic system will respond by releasing the neurotransmitter norepinephrine. The substantial work describing the body's response to stressful situations was done by Selye (1952) and is described as the General Adaptation Syndrome.

In times of stress, norepinephrine release will cause an increase in both heart rate and strength of cardiac contraction, thus providing additional blood flow to vital tissues. It will also increase contraction of smooth muscle in blood vessel walls, thus increasing blood return to the heart. The respiratory rate will also be increased. These responses result in the ability of a person to move rapidly from danger.

The function of the parasympathetic subsystem is to conserve body energy and to act in opposition to the sympathetic subdivision. The balance between these two subdivisions ultimately establishes the final output of many body organs and regulates functions such as digestion, elimination, respiration, and heart rate. The neurotransmitter acetylcholine is the output regulator for the parasympathetic division.

Although the autonomic system has in the past not been considered part of the cognator regulation (conscious control), some research indicates otherwise. The theory of biofeedback suggests that a person can be taught to control many autonomic functions. Wiener (1948) postulated that the body uses information about past performance as a control mechanism for further action. Miller (1969) expanded on Wiener's work and found that through conditioning exercises a person could control many of the autonomic functions. Cohen and Obrist (1975) further elucidated this idea through their work on teaching subjects to control blood pressure and heart rate through biofeedback conditioning.

Endocrine Subsystem

Another regulatory control system is that of the endocrine glands. This subsystem provides control through glands such as the pituitary, adrenal, thyroid, and sex glands, and the islets of Langerhans in the pancreas. Endocrine control is achieved by production of chemical transmitters called hormones. These chemical transmitters released by endocrine glands are transported through the vascular system to other cells in the body at considerable distance from the source. Hormonal controls exist for such diverse functions as growth, reproduction, body metabolism, and the maintenance of the internal milieu. Organ response to endocrine control is slower than that arising from the nervous system control.

The endocrine subsystem operates under tremendously complex controls using primarily negative feedback. An example of a negative feedback system is that of glucocorticoid regulation. The blood levels of the hormone cortisol, a glucocorticoid, are carefully regulated by a complex system including the hypothalamus, anterior pituitary, and adrenal glands. If blood levels of cortisol fall below normal, sensors in the hypothalamus react to the low cortisol level and stimulate the secretion of cortisol releasing hormone (CRH) (Guillemin & Rosenburg, 1955; Saffran & Schally, 1955). CRH then stimulates the anterior pituitary to release adrenocorticotrophic hormone (ACTH). ACTH, on reaching the adrenal gland, is the input signal for secretion of additional cortisol. The rise of cortisol in the blood will then "shut off" release of CRH in the hypothalamus and stop the process. Thus, as blood levels of cortisol increase, a negative feedback mechanism shuts off further release from the adrenal gland. This negative feedback schema is depicted in Figure 5.1.

The secretion of cortisol, however, is controlled by other factors as well. Cortisol release follows a circadian rhythm cycling over a 24-hour period. Halberg (1961) and Bartter, Delea, and Halberg (1961) mapped the levels of cortisol in the blood during a 24-hour period. They found that increased cortisol release occurs just prior to one's normal awakening time. It is believed

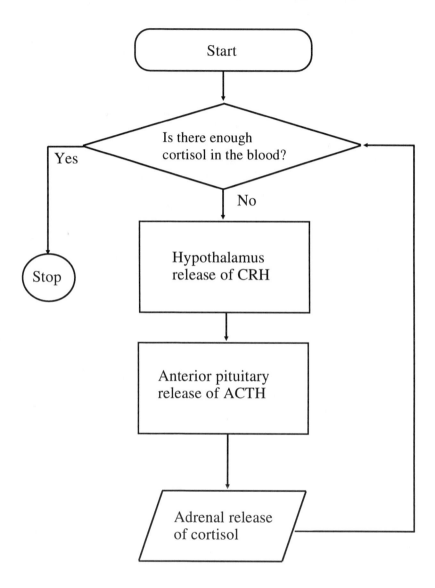

Figure 5.1. Negative Feedback Control Glucocorticoid Regulation

that this release of cortisol is what provides the person with the energy to start the new day. Thus, more than just a negative feedback mechanism is controlling cortisol release. The "set point" for the amount of cortisol released fluctuates during a 24-hour period.

Cortisol release is also affected by stress. Selye's (1952) General Adaptation Syndrome theory elucidated this pattern. Studies have shown that cortisol release is directly increased as a person is subjected to stress.

Other similar control mechanisms exist for each of the body's hormones. The amount of a hormone in the blood acts as the input signal to a target gland to either increase or decrease its activity. For example, cortisol released into the bloodstream will affect many tissues, causing increased production of glucose in the liver, reduced cellular utilization of glucose, and increased mobilization of amino acids from tissues. The net effect of these actions is believed to provide needed chemicals for immediate energy and tissue repair. Another effect of increased cortisol is a decreased response to foreign tissues, thus inhibiting the immune system function.

This example illustrates the application of systems in the hormonal system. Each endocrine gland is stimulated by an input signal that is peculiar to itself. Its throughput is to produce other chemicals that again are specific to the gland. Finally, each endocrine gland releases its product output that serves to control some other tissue in the body.

Immune Subsystem

The third portion of the regulatory subsystem is the immune system. Its function is to provide protection from bacteria, viruses, and other substances that enter the body from the external environment. It also provides protection from tissues that are different from self. Each cell within the body carries a "marker" that identifies itself as part of that person. Cells that do not belong to the person stimulate a response that identifies the cell as foreign to self. This identification initiates a chain of reactions that result in destruction of foreign substances.

The type of cell that controls this recognition system is one of the white blood cells, the T lymphocytes. When a foreign substance finds entry into the internal environment, the lymphocyte initiates the production of antibodies that prepare the foreign substance for destruction. It also initiates the activity of two other types of white cells, the polymorphonuclear leukocytes and the monocytes, which then destroy the damaged foreign substance.

The advent of the human immunodeficiency virus (HIV) that has produced the acquired immunodeficiency disease (AIDS) epidemic attests to the importance of this system as a regulatory subsystem. The HIV virus invades the lymphocyte cell, where it can remain hidden for long periods of time and eventually destroy it. It has been demonstrated that when the number of CD4 T lymphocytes, the specific type of lymphocyte invaded by the HIV virus, falls to less than $150/mm^3$, a strong correlation with the progression of AIDS exists (Leen & Brettle, 1992).

The reduction of CD4 T lymphocytes results in loss of a number of the immune subsystem responses. Leen and Brettle (1992) outline the major immunological changes that occur with the reduction in the CD4 T cell counts. Pathogens such as latent viruses, fungi, and protozoa that are normally held in check by cell-mediated immunological responses proliferate. The normal inflammatory response to foreign materials, mediated by macrophage cells, does not occur. There is also a decrease in antibody production in response to a foreign substance. Young children who have not yet developed memory cells for foreign substances will have significant reduction in the gamma globulin levels and consequently reduced levels of antibodies.

Circulatory and Respiratory Subsystem

A major function of the circulatory and respiratory subsystem is to supply cells with the materials needed to carry out energy production through oxidation. A corollary of this function is the need to transport these materials throughout the body.

Oxygen is a vital component of tissue metabolism. To supply the needed oxygen, respiratory ventilation must be adequate to provide the arterial blood with oxygen. The cardiovascular system must also produce a cardiac output sufficient to perfuse all the body tissues. Regulation of both processes must be continuous to maintain cell function. The cardiovascular system is also important in transporting other materials such as nutrients, vitamins, and minerals to the cells and waste products from tissue activity.

Respiratory ventilation begins when gas moves into the lungs through coordinated action of the respiratory muscles responding to controls of the nervous system. Oxygen selectively diffuses across the alveolar membrane into the blood capillary and attaches to hemoglobin molecules contained within the red blood cell. Oxygenated blood returns from the pulmonary circulation to the left side of the heart and from there is circulated throughout the body. At the cellular level, the oxygen diffuses into cells and is used in energy production. The waste product from energy production, carbon dioxide, in turn is then picked up from the cell by the carbaminohemoglobin molecule as well as through simple diffusion in the blood and carried back to the lungs through the cardiovascular system.

Carbon dioxide is the primary input signal to the central nervous subsystem regulating ventilation (Guyton, 1991). As the carbon dioxide level rises, the throughput response in the central nervous system is to generate impulses that stimulate respiratory muscles. The output response is to increase the depth and rate of the respiratory effort, thus increasing the amount of oxygen brought into the lungs and available for transport to cells. The system control mechanisms for maintaining adequate tissue oxygenation are extremely responsive to minor alterations in the input signal and are highly effective in maintaining an adequate cellular oxygenation.

There are several control mechanisms that regulate cardiovascular transport of the materials throughout the body. The first is related to the filling pressure of the heart at the end of diastole, the preload. According to Starling's (1918) law, the amount of force with which the heart will contract is directly related to the amount of stretch on the fibers. As the heart muscle fibers are stretched by increased fluid volume in the chambers, the force of contraction is increased and the amount of volume forced out of the heart in systole is increased.

The second regulatory mechanism is control of tone in the blood vessels. The nervous system, through the action of the sympathetic fibers, can initiate an increase or decrease in the size of the blood vessels, primarily in the venous circulation. This phenomenon was first identified by Brown-Sequard (1858) and Bernard (1851, 1852). If these vessels dilate, an enormous amount of fluid is retained in the venous system. This will decrease the preload of the heart and result in reduced cardiac output. This, in turn, lowers the pressure in the arterial system, the blood pressure, and reduces circulation to the tissue cells.

The regulation of cellular supply of oxygen and nutrients is truly an interactional model in which the respiratory and cardiac tissues receive constant input from within the body's internal cellular environment. Their output allows for moment-by-moment adaptation to meet cellular tissue needs.

Musculoskeletal Subsystem

The musculoskeletal subsystem has several major functions important to body system homeostasis. Muscle cell contraction provides the power for movement of the body in response to internal needs as well as needs stimulated by the external environment. The skeletal system, consisting of the bony structure, provides the support to the body to allow for movement through the environment. It also has functions that respond to the internal environment. For example, the red blood cells needed for oxygen transport arise within the central portion of the bone. The white blood cells needed for protection of the internal environment also originate there.

Muscle cells are of three types. The first is skeletal muscle, which provides for locomotion of the body by direct inputs from the central nervous system. In addition, smooth muscle cells in a number of body organs provide for movement of materials through these organs. For example, in the walls of the tissues making up the gastrointestinal tract, the smooth-muscle cell contraction moves the materials taken into the body throughout the tract. Without this function, one would not be able to transport the food materials ingested to the location where they can be prepared for assimilation into the internal body. Smooth muscle is also found in many other organs such as the bladder, ureters, and reproductive organs. The third type of muscle is found in the heart. It provides the pumping force for the transport of materials throughout the circulatory system. The smooth and cardiac muscles are not part of the musculoskeletal system. They are mentioned here because their contractile mechanism is similar to that of skeletal muscle.

The musculoskeletal subsystem is in constant contact with the larger external environment through inputs from the central nervous system. The throughput activity of muscle contraction is a chemical reaction between a protein, actomyosin, found within each muscle cell, and the energy source adenosine triphosphate (ATP). This reaction has been described by Szent-Gyorgyi (1968). As the result of the actomyosin-ATP interaction, the muscle fiber shortens and the output, movement, occurs.

Gastrointestinal Subsystem

The concepts related to the gastrointestinal system include the regulation of nutrient intake and preparation of nutrients for absorption into the blood. The elimination of wastes from cell metabolism is also important. Input signals for the ingestion of food are believed to be multiple. One fairly simple input signal is that of the sense of smell. The control mechanisms that regulate the amount of food intake are described in theories encompassing control mechanisms originating in the peripheral tissue and within the central nervous system. There appears to be considerable redundancy in these control mechanisms, perhaps in an effort to fine-tune the food intake.

The peripheral theories for the control of hunger and satiety include signals arising in the adrenal gland, liver, pancreas, and gastrointestinal tract (Chafetz, 1990; Cooper & Liebman, 1992). Nutrients and hormones present in the blood are important to coordination of eating behaviors (Leibowitz, 1991). Changes in response to these input signals can occur across the daily cycle as well as from day to day and even from one season to the next.

Within the central nervous system, the hypothalamus is believed to be the control center for satiety. Several neurochemicals and neuroendocrine signals are used to modulate the desire for food intake (Leibowitz, 1992). These act through the feeding and satiety centers within the hypothalamus of the brain that influence a person to start and stop eating. There two areas, the lateral hypothalamus (LH) and the medial hypothalamus (MH), respond to stimuli arising from other areas of the body. When the LH is stimulated or the MH is depressed, a person will increase food intake. When just the opposite occurs, MH stimulation or LH depression, food intake is reduced. These relationships are shown in Table 5.1.

Table 5.1   Effect of Stimulation of Hypothalamus on Food Intake[a]

	Medial Hypothalamus	Lateral Hypothalamus
Stimulation	–	+
Depression	+	–

a. + Indicates an increase in food intake; – indicates a decrease in food intake.

Once the food is ingested, the throughput mechanisms become operational. These include both preparation of the food for absorption and movement of the food throughout the body. The smooth muscle cells of the gastrointestinal system are responsible for its movement. Enzymes within this system are used to break down the complex carbohydrates, proteins, and fats into simple products that can be absorbed through the membranes of the gastrointestinal tract into the blood vessels. From there, these products are moved to the liver. The liver serves as a giant control system that is in constant contact with the internal environment and can keep the supply of needed nutrients readily available for cell usage (Newsholme & Start, 1973).

In the fed state, the hormone insulin is present in large amounts. Insulin accelerates the deposition of glycogen in the liver and to a lesser degree in skeletal muscle cells. In the fasting state, however, when blood insulin levels are low and another hormone, glucagon, is high, the opposite occurs. Glucagon stimulates the breakdown of liver glycogen into glucose that is then released into the blood, increasing the blood glucose levels again to a normal state. Another source of glucose formation is protein stored in the body tissues. After depletion of liver glycogen stores, 60% of the amino acids can be readily converted to glucose. The hormone cortisol aids glucagon in this process. A third source of potential glucose is lipid stores in the body. An enzyme, lipase, is activated in adipose tissue and serves to break down fats into free fatty acids and glycerol. Both these molecules can be oxidized to carbon dioxide, water, and energy.

The net effect of all these activities is the provision of nutrients to body cells as needed. A complex regulatory system operates to control the supply of nutrients so that a person's feeding periods can be at scattered intervals. Of course, any interference with this regulatory control will result in inadequate nutrition.

Renal Subsystem

The renal subsystem consists of two kidneys that receive arterial blood directly from the heart. Considering the function of the renal system from a systems point of view, the blood can be viewed as the input signal. The throughput mechanisms act on this blood as it moves through the nephron unit within the kidney to regulate both fluid and electrolyte balance within the body and to remove body waste products. Two such waste products, urea and creatinine, from protein metabolism are excreted into the output of the system, urine. Other materials, including glucose and the electrolytes such as sodium, potassium, calcium, and phosphorus, are selectively removed so that the blood levels of these chemicals are maintained within narrow limits. The body's acid-base balance is also maintained by selectively retaining or excreting hydrogen ions according to the needs of the cells. The materials not required by the body cells are excreted into the urine and removed from the body.

At the same time, the nephron unit is balancing the body's need for water by either retaining most of the water when fluid volume is low or by excreting large amounts of water if the fluid volume is in excess. The control regulation for maintaining this delicate balance is through the countercurrent mechanism (Berliner, Levinsky, Davidson, & Eden, 1958). This theory gives an explanation of how the body is able to produce a waste product, urine, of much greater concentration than the fluid in the vascular system. The filtrate that moves through into the nephron unit from Bowman's capsule is gradually concentrated by the removal of water from the filtrate through an osmotic gradient. This gradient produces an exchange of water, sodium, and other electrolytes to produce the final urine concentration. At the same time, some waste products, such as urea and creatinine, are not able to be transported through tubule membranes and remain in the portion that will become the urine. The final concentrating effect is regulated by the hormones aldosterone and antidiuretic hormone. The final urine concentration is such that 99% of the water and electrolytes entering the filtrate is returned to the vascular bed. In this way, the renal system is able to maintain a stable fluid and electrolyte balance in harmony with the efficient functioning of cells.

Although the renal system function is directly dependent on a continuous blood supply to provide the filtrate through the nephron unit to carry out this regulation, the kidney in turn can have a regulatory effect on the function of the heart. Normal blood flow to the kidney is about 160 L of blood per day or about 160 ml per minute. If this flow drops, a hormone, renin, is secreted (Goldblatt, Lynch, Hanzal, & Summerville, 1934; Tigerstedt & Bergman, 1898). The renin causes release of angiotensinogen, a precursor of the hormone angiotensin. In the lungs, this precursor is altered by angiotensin-converting enzyme to the active hormone, angiotensin. One action of angiotensin is to increase systemic vasoconstriction (Braun-Menendez, Fasciolo, LeLoir, & Munoz, 1940; Page & Helmer, 1940). Through the increased vasoconstriction, the systolic blood pressure is increased (Skeggs, Kahn, & Shumway, 1951). Angiotensin also stimulates the release of aldosterone from

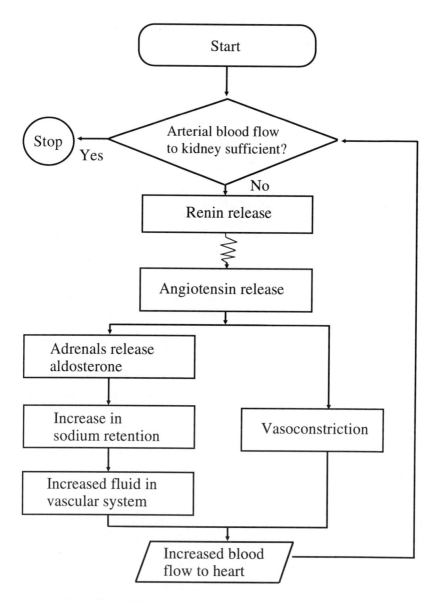

Figure 5.2. RAA System

the adrenal gland. The effect of aldosterone on the kidney tubule is to selectively return more sodium to the vascular blood. The net effect of both these actions is to increase the amount of fluid returned to the heart, and the cardiac output is increased. These interactions are shown in Figure 5.2. Using all these mechanisms, one can see that the kidney has some ability to control its own destiny.

Reproductive Subsystem

Finally, the reproductive subsystem is vital to perpetuation of the species by a unique series of activities. It is regulated by hormonal control operating under a negative feedback system similar to that discussed previously. The system carries out differing functions at three distinct periods in a person's life. In the embryo phase, differentiation of male or female characteristics is controlled. At puberty, growth of the genital organs is regulated. Finally, in adulthood, production of eggs within the female ovary and sperm within the male testis is regulated.

The center for the control mechanisms in both sexes is within the hypothalamus through regulation of the gonadotropin releasing hormone. This hormone is the input stimulus for all the activities of the reproductive system. It travels to the anterior pituitary gland where it stimulates the release of follicle stimulating hormone (FSH) and luteinizing hormone. The target tissues for these two hormones differ between the sexes.

In the male, FSH controls the production of spermatogenesis. Luteinizing hormone controls the release of testosterone, which is necessary for the maturation of the spermatozoa produced. Testosterone is also the hormone necessary for the differentiation of the internal and external male genitalia in the embryo and the male characteristics developed at puberty.

In the female, luteinizing hormone regulates the production of progesterone that is important to the development of uterine endometrium and breast tissue. FSH initiates release of estrogen that in the adult regulates ovulation, implantation of a fertilized egg, pregnancy, parturition, and lactation. It also is the control hormone for female sex organ development in the fetus and the female sex changes that occur at puberty.

The control of all these functions is vital to the full development of both male and female characteristics and activities. The output of this entire system makes regeneration of the race possible. Thus, it can be considered the vital link between one generation and the next.

Clinical Example

In addition to considering the separate activities of the biological subsystems, nurses should pay some attention to the interrelatedness of these subsystems functioning cooperatively to maintain an internal cellular environment conducive to life. One example of how these subsystems function together is the body's adaptation to an external environmental change such as moving from a sea level location to a high-altitude location.

The person who has adapted to life in the atmospheric conditions at sea level will find himself or herself at a severe oxygen deficit when faced with a high-altitude atmosphere. At this level, the total oxygen available in the atmosphere is considerably less than at sea level. During the period of adaptation to this environmental stressor, the person will display symptoms of fatigue, shortness of breath, headache, and other indicators of cellular oxygen deficit. The person's response to the situational stressor of decreased cellular oxygen is to make a biological adaptation.

Thus, it can be seen that the biological subsystems not only carry out their unique functions but are also in constant interaction with other subsystems to maintain the body's homeostasis. Continuous response to input signals in each of the subsystems provides the necessary alterations so that the person can respond to all the challenges of living in a constantly changing external environment. The following clinical example illustrates the adaptation to an environmental stressor.

Focus of Situational Interaction

Tim Jones was referred to the Nurse Case Manager (NCM), Debra, by his physician because he is experiencing shortness of breath on exertion. He is confused by his symptoms because he never had any problem with his breathing as an adult.

Assessment

Developmental Environment

Biological Subsystem

While hiking, Tim became short of breath and was taken to the local emergency department. He was treated with oxygen and bron-chodilators and was released home. Since moving to Northern Arizona, Tim has experienced shortness of breath while walking uphill or lifting boxes in his store. He also complains of a constant low-grade headache and fatigue. He has a history of childhood asthma.

Psychosocial Subsystem

Tim Jones, a 45-year-old single man, had decided to make a mid-career change. He moved from the Southern California basin, where he had been an advertising executive, to Northern Arizona, where he bought a small business—an art framing shop. His motivation in making this move was to find a place to live where he would be able to pursue his outdoor hobbies of hiking and cross-country skiing in a more relaxed atmosphere.

Spiritual Subsystem

Nothing is known about his religious preferences or activities.

Situational Environment

Intrasystem Analysis of Client

Detector Function. Tim Jones has little understanding of the physiology of respiration. He does not understand the oxygen deficit he is experiencing and the changes his body must make to adapt to an increase in altitude. He does not know how to interpret his lab tests.

Selector Function. Tim values good health and the ability to pursue an active outdoor life. He is interested in improving his quality of life.

Effector Function. Tim has attempted to carry on life as usual although he is aware that any exertion leaves him short of breath.

Intrasystem Analysis of Nurse Case Manager

Detector Function. The NCM understands that the biological subsystem will quickly respond to an environmental change—that cells within the renal system will detect the oxygen deficit that will cause an outpouring of the hormone erythropoietin, which stimulates bone marrow to increase the release of red blood cells. With an increase in the number of red blood cells, the oxygen-carrying capacity of the blood will be increased so that the limited amount of oxygen in the alveoli will be picked up more efficiently.

Selector Function. The NCM values quality of life both for herself and for her clients. She enjoys helping clients make life-changing adaptations.

Effector Function. The NCM set up the clinical tests ordered by the physician to assess Tim's respiratory status.

Analysis

Identification of Stressors

Recent move from sea level to 7,000 feet
Childhood asthma

Coping Resources

Wants to pursue outdoor hobbies
Has HMO insurance
Has a new business that interests him

Scoring on Situational Sense of Coherence

Comprehensibility = 1. He has little knowledge of what is happening in his body.

Meaningfulness = 3. He is highly motivated to improve his health status and exercise tolerance.

Manageability = 3. He has the resources necessary to assist his adaptation to the increased altitude.

Nursing Diagnosis

Activity intolerance related to decreased tissue oxygenation resulting from the environmental change.

Nursing Goals/Client Goals

1. To assess Tim's respiratory status
2. To increase Tim's understanding of respiratory physiology and needed lifestyle changes
3. To assist Tim to improve his exercise tolerance

Intersystem Interaction

Communication of Information

Information about changes in tissue oxygenation because of the elevation change and the necessary lifestyle changes needed to improve oxygenation were given. Specifically, the NCM instructed him about breathing techniques that will increase gas exchange in the lungs. She also emphasized the need for periods of rest interspersed with activities and about dietary changes to increase available iron for the production of additional red blood cells. She assured him that, in time, his body would be successful in making the adaptation to the changed environment.

Negotiation of Values

Both NCM and client were in agreement about the life changes needed. The NCM expressed appreciation for the value system expressed by Tim.

Organization of Behaviors: Mutual Plan of Care

Laboratory test were completed and interpreted. An iron-rich diet was started. Breathing exercises were learned. A schedule of rest and activity was developed.

Evaluation of the Plan: Rescoring on SSOC

Comprehensibility = 3. Tim understood all explanations and was able to incorporate them into his lifestyle.

Meaningfulness = 3. Tim continues to feel the lifestyle changes he has begun are important to his health status.

Manageability = 3. Tim continues to have the resources necessary to carry out his new program.

References

Asratian, E. A. (1965). *Compensatory adaptations, reflex activity, and the brain.* New York: Pergamon.

Bartter, F. C., Delea, C. S., & Halberg, F. (1961). A map of blood and urinary changes related to circadian variation in adrenal cortical function in normal subjects. Symposium on rhythmic functions in the living system. *Annals, New York Academy of Science, 98,* 753.

Berliner, R. W., Levinsky, N. G., Davidson, D. G., & Eden, M. (1958). Dilution and concentration of the urine and action of antidiuretic hormone. *American Journal of Medicine, 24,* 730.

Bernard, C. (1851). Influence du grand sympathique sur la sensibilite et sur la calorification. *CR Soc Biol (Paris), 3,* 163-164.

Bernard, C. (1852). Sur les effete de la section de la position cephalique du grand sympathetique. *C R Soc Biol (Paris), 25,* 168-170.

Bertalanffy, L. von (1968). *General systems theory.* Harmondsworth, UK: Penguin.

Braun-Menendez, E., Fasciolo, J. C., LeLoir, L. F., & Munoz, J. M. (1940). The substance causing renal hypertension. *Journal of Physiology, 98,* 283-298.

Brown-Sequard, C. E. (1858). Sur la sensibilité tactile et sur un moyen de la mesurer dans l'anesthésie et l'hyperesthésie. Augmentation de la finesse du sens du lieu par suite de l'hyperesthésie. *Journal Physiol (Paris), 1,* 344-348.

Casey, K. L., & Melzack, R. (1967). Neural mechanisms of pain: A conceptual model. In E. L. Way (Ed.), *New concepts of pain and its clinical management.* Philadelphia: F. A. Davis.

Chafetz, M. D. (1990). *Nutrition and neurotransmitters.* Englewood Cliffs, NJ: Prentice Hall.

Cohen, D. H., & Obrist, P. A. (1975). Interactions between behavior and the cardiovascular system. *Circulation Research, 37,* 693-706.

Cooper, S. J., & Liebman, J. (1992). *The neuropharmacology of appetite.* Oxford, UK: Oxford University Press.

Goldblatt, H., Lynch, J., Hanzal, R. F., & Summerville, W. W. (1934). Studies on experimental hypertension: I. The production of persistent elevation of systolic blood pressure by means of renal ischemia. *Journal of Experimental Medicine, 59,* 347-379.

Guillemin, R., & Rosenburg, B. (1955). Humoral hypothalamic control of anterior pituitary: A study with combined tissue cultures. *Endocrinology, 57,* 599-607.

Guyton, A. (1991). *Textbook of medical physiology.* Philadelphia: W. B. Saunders.

Halberg, F. (1961). *Adrenal cycle.* Paper presented at the 39th Ross Conference Pediatric Research, Minneapolis, MN.

Leen, C. L., & Brettle, R. P. (1992). Natural history of HIV infection. In A. G. Bird (Ed.), *Immunology of HIV infection.* Dordrecht, The Netherlands: Kluwer Academic.

Leibowitz, S. F. (1991). *Obesity and cachexia.* N. J. Rothwell & M. J. Stock (Eds.), (pp. 33-48). New York: John Wiley.

Meinhart, N. T., & McCaffery, M. (1983). *Pain* (pp. 58-60). Norwalk, CT: Appleton-Century-Crofts.

Melzack, R., & Wall, P. D. (1965). Pain mechanisms: A new theory. *Science, 150,* 971-979.

Miller, N. E. (1969). Learning of visceral and glandular responses. *Science, 163,* 434-445.

Newsholme, E. A., & Start, C. (1973). *Regulation in metabolism.* London: Wiley.

Page, I. H., & Helmer, O. M. (1940). A crystalline pressor substance (angiotensin) resulting from the reaction between renin and renin activator. *Journal of Experimental Medicine, 71,* 29-42.

Saffran, M., & Schally, A. V. (1955). Stimulation of the release of corticotrophin from the adenohypophysis by a neurohypophyseal factor. *Endocrinology, 57,* 439-444.

Selye, H. (1952). *The story of the adaptation syndrome.* Montreal: Acta.

Skeggs, L. T., Kahn, J. R., & Shumway, N. P. (1951). The isolation of hypertensin from the circulating blood of normal dogs with experimental renal hypertension by dialysis in an artificial kidney. *Circulation, 3,* 384-389.

Starling, E. H. (1918). *The Linacre lecture on the law of the heart.* London: Longmans, Green.

Szent-Gyorgyi, A. (1968). The role of actin-myosin interaction in contraction. *Symposium of the Society for Experimental Biology, 22*(17).

Tigerstedt, R., & Bergman, P. G. (1898). Niere und Kreislauf. *Scandinavian Archives of Physiology, 8,* 223-271.

Wiener, N. (1948). *Cybernetics or control and communication in the animal and the machine.* New York: John Wiley.

6

Theoretical Foundations for the Psychosocial Domain of Human Experience

Margaret England

Input Into the Psychosocial Subsystem
Transformations of the Psychosocial Subsystem
Outputs of the Psychosocial Subsystem
Clinical Example

The psychosocial subsystem of the person is the social self of the person that relates to self and others. It includes the cognitive subsystem, which processes information; the affective subsystem, which displays feeling and emotions mediated through values; and the interactional subsystem, which organizes behavior in established patterns. The unique configuration of the psychosocial domain of human experience emerges from experience within and across all domains of experience. The interplay of experiences within other domains of experience provides feedback to the affected domain so that it can further pattern itself. Concepts useful in understanding the psychosocial subsystem are derived from theories from the biopsychosocial sciences that describe the coherent patterning of psychosocial human experiences and responses. Key concepts within the theories are discussed in this chapter as they relate to input into the psychosocial subsystem, to transformations of experience within the subsystem, and to behavioral manifestations. Their implications for clinical practice are presented and related to relevant nursing knowledge, values, and behaviors.

Input Into the Psychosocial Subsystem

Input into the psychosocial subsystem arises from the external and internal environments of both the client and the nurse. The external environment, denoted as the situational environment in the Intersystem Model, consists of events and the communications of the client and the nurse as they both respond to the client's health situation. The internal environment, in contrast, is who the person is—that is, a unified and dynamic constellation of biopsychosocial experiences. Maturana and Varela (1987) describe this unique sense of self as the "niche" of the person. The client and the nurse have grown up in separate, sometimes very different, developmental environments, and those environments have made them the persons they are. Differing constellations of background experiences will prompt the client and nurse to respond to one another differently. Previous constellations of experience and current interactions of the client and nurse will produce still other unique constellations of experience, changing who they are as persons.

Nurses can zero in on the domain of psychosocial experience because, like other human beings, they have the capacity to integrate human experiences within that domain. Nurses can fulfill their responsibility through alliances with the personhood of the individual and their interpersonal network.

Transformations of the Psychosocial Subsystem

The Intersystem Model places much importance on the coherent patterning of human responses. According to the model, the coherent patterning of human responses occurs because of transformations that take place within and across biological, psychosocial, and spiritual domains of human experience. Unique transformations taking place within one domain of human experience are both antecedents and consequences of unique transformations occurring within and across other domains of experience.

Transformations of human experience within the psychosocial domain can be subsumed within the cognitive, affective, and interactional subsystems of the psychosocial subsystem. Key concepts relating to transformation, therefore, are grouped according to how they relate to cognition, affect, and interaction.

Cognitive Subsystem

Evidence of a person's ability to transform cognitions within the psychosocial domain of human experience can be explained with concepts such as perception, stress, and perceived health. The concepts refer to ways in which individuals develop constructs about themselves and other people. Nurses can use these concepts to help explain an individual's subjective human responses and experiences.

Perception

Paramount to understanding a person's response to a health situation is an understanding of how situations are perceived. A person takes in information about a situation through the detector function. The person transforms that information according to how he or she personally filters the information through the selector function. The person then records the information in memory.

Persons vary in their ability to process information. Low intellectual level or inexperience may prevent some individuals from developing categories within which to order information from the environment. An undifferentiated self, the inability to distinguish the self from other objects, may prevent other individuals from developing holistic images of a situation.

Loevinger (1976) theorized that a person mentally creates and restructures knowledge about the self and world continuously and, in the process, the person makes finer and finer discriminations about elements of the self and the world. When this process is not complete, perceptions of situations will be incomplete. Perceptions can also be distorted by an unwillingness to perceive the situation for what it is. For example, an adult offspring may comment on a sudden change in the mental status of a parent who, in actuality, had been confused for a long time. In this case, the adult offspring may have been unwilling to process behavioral inputs accurately because of his or her need to view the parent as competent.

Stress

Perceptions of a health situation may also be influenced by the way in which persons appraise and respond to the demands of a situation. Kuhn (1974) defined stress as "an act or situation that . . . generates a desire that it be removed" (p. 195). Lazarus and Folkman's (1984) theory of stress, appraisal, and coping and Apter's (1982, 1984) reversal theory are two perspectives that can help explain the extent to which individuals can and do adapt to stressful life experiences.

Appraisal and Coping. According to Lazarus and colleagues, stress is an event in which environmental or internal demands or both exceed the adaptive resources of an individual, social system, or tissue system (Arnold, 1960a, 1960b, 1970; Lazarus, 1966; Lazarus & Folkman, 1984; Lazarus & Launier, 1978). Stressors are internal or external events or both that impose adaptive requirements on an individual or social system. For Lazarus, a stimulus becomes a stressor when it produces a distressing behavioral or physiological response. Similarly, a human response is stressful when an individual perceives demand, harm, threat, loads, or burdens that have a negative impact on his or her well-being.

Psychological stress consists of a unique relationship between the individual and the environment that hinges on the concept of cognitive appraisal. Cognitive appraisal is a mental process of placing an event in a valuative category related either to its significance for an individual's well-being (primary appraisal) or to an individual's available coping resources and options (secondary appraisal). Appraisal is an automatic intuitive process that has two functions. The first function of appraisal is the determination of why and to what extent a particular situation is threatening. The second function is the determination of why and to what extent coping behavior is successful.

Individual, situational, temporal, and other factors influence or moderate appraisal and coping processes in a threatening encounter in one of three ways. The first way of moderating appraisal and coping is through determination of what is salient for well-being in a threatening situation. The second way of moderating appraisal and coping is through understanding of the situation and potential consequences. The final medium of influence is the provision of a basis for evaluating success in the situation.

Personal factors that influence appraisal and coping processes include preferences or values, beliefs, personal vulnerability, and depth of awareness. Preferences or values are an expression of what is important to the individual and the choices he or she makes. Beliefs are personally formed or shared notions or both about reality that influence how an individual may evaluate the demands of a situation and how he or she will manage the situation. Existential beliefs enable the individual to create meaning and maintain hope in the situation. Personal vulnerability refers to an actual or perceived lack of sufficient resources for managing an encounter. Depth of awareness refers to an individual's level of awareness of the significance of an event for his or her well-being.

Situational factors, such as novelty, predictability, and event uncertainty, influence appraisals in a threatening encounter as well as the significance of the encounter for well-being. Temporal factors of imminence, duration, and timing of an event can also influence appraisal. Ambiguity and uncertainty are other characteristics that can influence the relationship between the individual and environment in a threatening encounter.

According to Lazarus and Folkman (1984), successful coping leads to short-term and long-term adaptation. In their view, there are three major outcomes of coping. These are social functioning, morale and life satisfaction, and somatic health. Social functioning is the effectiveness with which an individual appraises and copes with the demands of everyday living. It includes effective functioning in interpersonal relationships and social roles. Morale is the positive and negative emotions an individual experiences during and after a threatening encounter. Morale incorporates quality of life and satisfaction with living. Somatic health refers to states of physiological change that accrue because of a threatening encounter.

Modulation of Stress. Individuals differ in the degree to which they are adversely affected by threatening encounters and negative life experiences. One principle for how well an individual can modulate his or her experiences in a threatening encounter relates to the availability and accessibility of personal resource characteristics (England, 1996). Resources can allow an individual to modulate strain associated with particular life situations. The term *modulation* refers to an ability to change or balance the impact of an event or encounter in a desired direction (England, 1993, p. 156). Individuals differ in the degree to which they are adversely affected by threatening encounters and negative life experiences. Those persons with a high sense of coherence will be able to manage stress and stay well to a higher degree than those with a low sense of coherence.

Reversal. In addition to resource characteristics, persons differ in their responses to stress because of temperament. According to reversal theory (Apter, 1982, 1991), there are two ways of experiencing stress: the telic and the paratelic modes of experience. A telic-dominant person prefers a quiet, serene environment in which to experience life, whereas the paratelic-dominant person prefers a more sensational environment. The telic-dominant person appears to be more ready to appraise situations as challenging. A telic-dominant person appears to feel and perform better under less stimulating, quieter, and more predictable circumstances than in highly charged situations. Conversely, the paratelic-dominant person appears to seek challenges because of his or her need for stimulation. Nevertheless, even the paratelic-dominant person has limits to his or her tolerance for stress experiences.

Perceived Health

Research shows that self-evaluations of health are not straightforward reflections of objective health status. Rather, they appear to represent complex cognitive maps that are influenced, in part, by perceptions admitted to consciousness, appraisals of a situation, and preferred responses to the challenge of the situation. In general, a self-report of health refers to an individual's interpretation of significant changes in health. An individual tends to use a sense of his or her own internal standard (Cantril, 1965) or a similarly aged reference group or both when making (reporting) an interpretation of his or her state of health (Schulz, Heckhausen, & Locher, 1991).

Cognitive-Illness Labeling-Reporting Process

Angel and Thoits (1987) proposed that individuals are likely to interpret changes in their internal state of health as they become aware of them. They call the process of labeling and reporting changes in health a cognitive-illness labeling and reporting process. The process occurs in three stages—the phases of monitoring, interpretation, and perceived seriousness. The first stage of the labeling-reporting process is referred to as the phase of monitoring. In theory, all individuals have a sense of the normalcy of their internal health state and become aware of changes from normal to not normal. Each individual has a unique threshold for determining changes in internal health state. Human responses indicative of significant negative changes in internal health state are called symptoms. When an individual becomes aware of symptoms, he or she begins to monitor the symptoms to determine deviations from normalcy.

The second stage of the illness labeling-reporting process is referred to as the phase of interpretation. An individual perceives changes in health state in a specific dimension. He or she then interprets, arrays, and labels the specific pattern of change within the dimension. An individual must reach and exceed his or her unique threshold for normalcy before he or she will be inclined to evaluate a negative change in health state. The third stage of the illness labeling-reporting process is referred to as the phase of perceived seriousness. An individual evaluates the seriousness of negative changes in internal health state and then translates this appraisal into a self-report of health. He or she then decides what might be done to alter a negative self-report if and when necessary.

Implications

Nurses need knowledge of perception, modes of responding to stress, and perceived health because these dimensions of cognition are a part of an individual's honest, cognitive-perceptual learning. That is, they represent ongoing "synthetic" transformations of an individual's level of awareness and ability to mold his or her human experiences and responses.

Nurses need to recognize and care for the unfolding of an individual's cognitive experience because it is the template for health promotion and healing. Not to recognize the cognitive-perceptual learning of the individual is to potentially undermine an individual's self-determining nature and nursing's societal mandate to nurture human potential. Nurses can fulfill their responsibility through nurturing and educative nursing actions.

Affective Subsystem

Evidence of a person's ability to transform feelings and other sensory experience within the psychosocial domain of human experience can be explained with concepts such as temperament, facial expression, and emotionality. These phenomena form the substrate for much of human experience. Nurses can use these concepts to help explain an individual's human responses and experiences.

Temperament

The concept of temperament refers to basic dispositions that underlie and modulate human responses and experiences with the environment (Goldsmith et al., 1987). There are three presuppositions commonly addressed in temperament theory. First, temperament is an inherently stable characteristic of the personality that has genetic, biological underpinnings. Second, temperament is a reflection of individual differences in tendencies to behave in certain ways. Third, direct expressions of temperament are subject to the influences of maturation, social context, and experience (see Table 6.1).

Domains of Temperament. Researchers identify four domains of temperament. The domains are activity, reactivity, emotionality, and sociability. Activity, the first domain of temperament, refers to behavioral evidence of the intensity, vigor, and pace of movement (Buss & Plomin, 1975). Activity can vary from lethargy to hypomanic pushes of energetic behavior. Expressions of activity include rate, amplitude, and duration of speaking and movement and the displacement of body movements in space and time. Reactivity, the second domain of temperament, refers to the excitability of human responses. Expressions of reactivity include approach-avoidance and withdrawal responses; thresholds for the initiation, maintenance, and termination of responses; attention and interest in stimuli; and persistence. Emotionality, the third domain of temperament, refers to a person's repertoire of negative and positive feeling states. Expressions of emotionality range from a stoic lack of reaction to intense reactions of being out of control. Sociability, the fourth domain of temperament, refers to a person's preference for being with others rather than being alone. Sociable individuals seek attention from others as well as the back-and-forth movement of social interaction. Expressions of sociability include attempts to initiate, maintain, and terminate social contact; time spent with others; social responsiveness; and reactions to being alone or isolated from others or both.

Temperament Style. According to Thomas and Chess (1977), temperament is a style of behavior that relates to the how of behavior rather than the why

Table 6.1   Definitions of Temperament and Their Underlying Assumptions

Reference	Definition	Assumptions
Buss and Plomin (1975); Plomin (1984)	A set of inherited personality traits that appear early in life	Biological origin Differentiate with maturation and experience
Thomas and Chess (1977, 1984, 1986); Thomas, Chess, and Birch (1968)	A style of behavior reflected in difficult, easy, or slow-to-warm up behaviors	Dynamic factor that mediates how the environment affects a person
Rothbart (1984); Rothbart and Derryberry (1981)	A set of individual differences in reactivity and self-regulation	Biological origin that is subject to developmental transitions Stable attribute of personality
Goldsmith (1984); Goldsmith and Gottesmann (1981)	Individual differences in the experience and expression of primary emotions and arousal	Subject to motivational influences that are independent of behavioral style

(motivation) or what (content) of behavior. In their view, temperament is a dynamic factor that shapes how the environment influences a person. Moreover, they link temperament with goodness of fit with the environment (Chess & Thomas, 1984, 1986, 1990). They refer to goodness of fit as an accord that is generated when opportunities, expectations, and demands of the environment cohere with a person's temperament.

Thomas and Chess (Chess & Thomas, 1984, 1986; Thomas & Chess, 1977; Thomas, Chess, & Birch, 1968) identified three dimensions of temperament. The dimensions are difficult, easy, and slow-to-warm-up temperaments. Difficult temperament refers to a specific cluster of attributes expressed as irregularity of biological functions, a tendency to withdraw in new situations, slow adaptability, intensity of mood, and relatively frequent episodes of negative mood. Easy temperament refers to the polar opposite of difficult temperament. Slow-to-warm-up temperament refers to a specific cluster of attributes expressed as shyness, lack of excitement, and relative unresponsiveness in new situations.

Temperamental Variability.   According to Rothbart (1984), temperament relates to inherently stable, individual differences in reactivity and self-regulation. Reactivity refers to the arousability of human experience. Self-regulation refers to the way in which individuals modulate that reactivity. Rothbart postulated that most individuals express variations in feelings of energy, interest, and affect. These tendencies are manifested in variable human response patterns of emotionality, activity, and attention.

Table 6.2   Linkages Between Action Tendency, Type of Goal, and Sanctioned Ritual
of Release for Various Coarse Emotions

Conceptual Framework for the Expression of Coarse Emotions			
Coarse Emotion	*Overt Action Tendency Sequence*	*Type of Goal*	*Socially Sanctioned Ritual of Release*
Joy	Smiling	Community	Celebration
Grief	Weeping	Separation	Mourning
Anger	Fighting	Integrity of self	Sports
Shame and embarrassment	Laughter	Idealized self	Comedy
Hunger	Ingestion	Satiation	Dining
Fear	Trembling	Safety	Protection

According to Rothbart (1984), temperamental variability influences an
individual's tolerance for stimulation and self-regulation. Variations in tolerance for stimulation will influence the kind of situation an individual places
himself or herself in as well as the kinds of strategies he or she uses for goal
attainment. The same variations in tolerance will also influence the way in
which an individual seeks out others for support and security.

Coarse Emotions

James (1890) outlined a theory of coarse emotions that relates to affective
transformations of human experience. According to the theory, coarse emotions are emotions that have a strong biological basis. Examples of coarse
emotions include anger, fear, pain, disgust, shame, grief, pleasure, joy, and
surprise. Coarse emotions are sometimes called primary emotions.

In theory, any coarse emotion can place the body in a state of arousal
(Scheff, 1984a, 1984b, 1985). In addition, every coarse emotion has a unique
goal and action tendency or emotional response cycle (Scheff, 1984b). An
emotional response cycle consists of generic sequences of concrete action (see
Table 6.2). In theory, proper completion of an emotional response cycle
cannot occur without catharsis.

Masters and Johnson (1966) describe four generic elements of an emotional
response cycle. They are stimulation, arousal, climax, and resolution. Stimulation
refers to mental images of human need to achieve the goal of an emotion. Arousal
refers to the lived experience of the emotion. Climax refers to the consummation
of the emotion. Resolution refers to general sense of relief, relaxation, and gradual
disappearance of the feeling state associated with the emotion.

Nichols and Zax (1977) provide a framework for identifying cognitive- and
somatic-emotional elements of catharsis. The cognitive-emotional elements

of catharsis refer to the reexperiencing of the contents of awareness by recalling and reliving an emotional event. By way of contrast, the somatic-emotional element of catharsis refers to the motoric discharge of emotion in expressive sounds and actions.

In general, the expression of a coarse emotion is associated with a characteristic goal, sequence of concrete actions or action tendency, and socially sanctioned ritual of release (see Table 6.2). In emotional arousal, the type of unique goal associated with a coarse emotion does not change. Nevertheless, a person's appreciation of the goal and the action tendency associated with the emotion may change because of maturation, social conditioning, and situational factors.

Every culture or society or both has culturally sanctioned rituals of release for the expression of coarse emotions. Socially sanctioned rituals for release are those rituals that evoke one or more permissible coarse emotions but only in a specific social context. The context for expression of a coarse emotion allows for catharsis and resolution of specific emotions. Effectively enacted rituals not only evoke unresolved coarse emotions but also evoke them in a context of safety and security.

Facial Expression

The concept *facial expression* can also be used to explain feeling states. Facial expressions are believed to be genetically and culturally programmed sequences of emotionality that are characteristic of all persons. Researchers such as Scheff (1985) agree that innate emotional activity occurs within the autonomic nervous system. That is, an individual literally is able to construct facial prototypes of an emotion muscle by muscle as the emotion unfolds. He or she is able to produce positive and negative emotions and make distinctions between emotions. In addition, the person learns to produce, avoid, or disguise facial expression in conformity to social norms because of social regulation and control.

Implications

Nurses need knowledge of temperament, the coarse emotions, and facial expression because these affective dimensions of human experience are a part of the biopsychosocial inheritance of the individual. That is, they represent natural, "God-given" transformation processes of the human system.

Nurses need to recognize and care for the natural unfolding of an individual's sensory experience and feelings because they are a salient barometer for health and healing. Not to recognize or respond to the biopsychosocial inheritance of the individual is to potentially "fool with Mother Nature" and nursing's societal mandate to nurture human potential.

Nurses bear responsibility for promoting the natural unfolding of the affective dimension of human experience. They can fulfill their responsibility by being fully aware of the purpose of sensory experience and feelings and by allowing for their proper and natural expression.

Interactional Subsystem

Evidence of a person's ability to transform interactions within the psychosocial domain of human experience can be explained with concepts such as relatedness, self-as-process, and self-system. These concepts are synthesized mental constructions concerning the self and other human systems. Nurses can use these concepts to help explain an individual's interpersonal human responses and experiences.

Relatedness

The concept *relatedness* appears to be an underpinning of an individual's need to be allied with other human systems. According to McMillan and Chavis (1986), relatedness relates to a sense of community with others. It is a feeling that the individual is connected with others in a meaningful way and can rely on them for fulfillment of his or her human needs.

Relatedness refers to a generalized capacity to integrate personal experience with other people given the reality of a present situation. It is an inner confidence that conditions the human spirit as an individual interacts with others. Shotter (1986) identified four basic presuppositions for relatedness. First, no human being is an island. Second, every human being needs a foundation as well as opportunities for human interaction. Third, confidence in self and others is a requisite for the proper modulation of human responses and experiences. Fourth, relatedness is essential for how an individual integrates his or her human experiences with others.

In summary, relatedness refers to how people integrate themselves experientially with others. Its health-producing qualities are embraced in the idea that people belong with one another and are cared about. Moreover, an individual with a sense of relatedness is able to believe in others and share a particular lifestyle or set of values and goals or both with them.

Self-as-Process

Symbolic interactionism is a perspective that relates to symbolic human experience. It derives from the social psychology of George Mead (Strauss, 1956) and addresses a basic question about the human condition: "What is it about the nature of human beings that permits them to develop a self-concept and relate with others?"

According to the perspective, human behavior is shaped by the meanings that individuals attach to social situations as a result of their interactions with others. An individual does not respond directly to the content and context of a situation. Instead, he or she assigns meaning to the social situation and responds on the basis of this meaning. Kuhn (1974) attributes this activity to the selector function of the person.

The concept *self-as-process* provides an understanding of how an individual might construct social situations in which to confirm his or her view of self and others. Mead (1934) defined the self as a process in the formulation and modification of social behavior. The construction of I and of Me are two phases in the self-as-process. The I, or initial phase of self-as-process, refers to the immediate, spontaneous, and impulsive aspects of behavior. The Me phase refers to the modulation of behavior following the immediate response. In the Me phase, an individual responds to self as an object in the situation via role taking.

According to Mead (1934), self-as-process is the shifting back and forth of I and Me. The role of Me presumes taking the point of view of the other in a situation. When an individual assumes the role of other, he or she becomes detached from his or her own behavior in the situation. The role of I presumes that an individual is not detached from his or her own behavior. When an individual assumes the role of I, he or she is both an actor and a participant in the situation.

Self-System

Van der Velde (1985) proposed that, with maturity, individuals become aware that they are being appraised by others at the same time that they themselves are appraising others. Individuals realize that they are not always certain that others interpret their appearance, movements, intentions, and communications in ways they would like. Individuals also learn that they must rely on the judgments of others even when they do not like them.

Sullivan (1953) presumed that the judgments of other people were crucial for the development of a self-system. He defined the self-system as an ongoing process of mental construction of an individual's ideas, feelings, and actions. According to him, the self-system emerges from interactions with others and becomes organized as a system of self-conceptions and patterns of behavior subject to further cultural molding.

Field (1979, pp. 33-36) described the development of a self-system as a series of six elements. First, an individual selectively hears appraisals of his or her self by others. Second, the individual attends to cues proffered by others that will help him or her master salient self-behaviors. Third, the individual incorporates others' reflected appraisals of his or her self as his or her own view of self. Fourth, the individual formats incorporated and reflected self-appraisals into corresponding personal actions that will enhance the meanings of such appraisals in specific situations. Fifth, the individual differentiates reflected self-appraisals into those that reflect his or her view of self (Me) and those that do not reflect such a view (Not Me). Finally, the individual constructs specific situations in which to confirm his or her particular view of self and others.

Implications

Nurses need knowledge of relatedness, self-as-process, and self-system because these dimensions of interaction are requisites for an individual's ability to form, sustain, transform, and terminate meaningful (and meaningless) relationships with others. They represent a psychosocial basis for a meaningful life and a peaceful death.

Nurses need to recognize and care for the unfolding of effective human interactions because they are templates for an individual's relatedness with other human systems. Not to recognize the interpersonal experience of the individual is to undermine the individual's need for relatedness with others and nursing's societal mandate to foster effective human interactions.

Nurses bear responsibility for the natural, learned unfolding of an individual's sense of relatedness and ability to communicate effectively with others. Nurses, through their use of self, can fulfill their responsibility by helping individuals recognize, transform, transcend, and release emotions and perceptions in an interpersonal context.

Outputs of the Psychosocial Subsystem

The Intersystem Model places much importance on the patterning of adaptive human responses and experiences in their proper context. According to the model, adaptive human responses are outcomes of transformational processes developed within the developmental environment of the human system. Adaptive human responses from the psychosocial domain can be represented as synthesized patterns of affects, cognitions, and interactions. They are associated with ongoing development of a situationally based sense of coherence.

Concepts relating to the patterning of adaptive human responses within the psychosocial domain of human experience include anxiety, crisis, and adaptation. Nurses can use these concepts to help explain the context and content of an individual's evolving human response trajectory. They can also use the concepts to help explain the content and goals of nursing care.

Anxiety

Anxiety is defined as a state of feeling apprehension, uneasiness, uncertainty, dread or any combination of the four stemming from a real or perceived threat whose actual source is unknown or not recognized (May, 1950). Theorists agree that anxiety consists of relatively consistent manifestations of cognitive and sensory perceptual human responses to threat.

There are four presuppositions commonly addressed in anxiety theory. First, anxiety is a universal experience that can be used to explain many human responses. Second, anxiety is a necessary life energy that motivates an individual to engage in activities of daily living and to make and achieve goals for living. Third, anxiety can be transmitted from one person to another through the mechanism of empathy. Fourth, anxiety permits people to make and survive change.

There are several patterns of distress related to anxiety that can be distinguished from one another. Among them are fear, primary and secondary anxiety, acute and chronic anxiety, anxiety syndrome, phobia, and posttraumatic stress disorder. Anxiety relates to a vaguely defined sense of danger, whereas fear relates to one or more specific dangers. Acute anxiety refers to present felt dread precipitated by an imminent loss or threat to security. Chronic anxiety refers to a relatively stable attitude of apprehension or to chronic overreaction or both to environmental stimuli and the circumstances of living.

Primary anxiety relates to interpersonal or intrapsychic conflicts, whereas secondary anxiety is caused by underlying biological processes. An anxiety syndrome refers to the experience of anxiety at multiple levels of experience, whereas a phobia is a generalized sense of anxiety at all levels of experience. Posttraumatic stress disorder refers to a set of symptoms that developed in relation to a lived experience of trauma.

There are two categories of threat linked with anxiety: threats to biological integrity and threats to the self-system. Threats to biological integrity include threats to the tendency of persons to achieve, maintain, and restore homeostasis through such processes as temperature control, maturation, and action sequences taken to resolve emotions and satisfy human needs. Threats to the self-system refer to threats that challenge an individual's established view of the self and the tendency to resist changes in self-perspective. Threats to self include unmet expectations and unmet needs for self-respect, prestige, status, and deference. Cues to the emergence and continuing presence of anxiety relate to biological arousal and threat to some value an individual holds dear to existence (May, 1950).

Levels of Anxiety

Peplau (1963) postulated that anxiety can be inferred on four levels of experience: mild, moderate, severe, and panic anxiety. Mild anxiety refers to an individual's ability to bring what he or she senses into sharp focus. A mildly anxious person is more attentive, able to grasp information quickly and clearly, and can solve problems effectively. The individual can observe, describe, analyze, and formulate meanings and relations. He or she can validate, test, integrate, and use the products of learning. The individual may experience mild discomfort such as restlessness or irritability.

Moderate anxiety refers to a narrowing of an individual's perceptual field. A moderately anxious person displays selective inattention and grasps less information than a less anxious person would in the same situation. The individual learns with difficulty and resolves problems better when helped or supported. The individual may experience arousal of the autonomic nervous system and may evidence such symptoms as increased respiration, heart rate and systolic blood pressure, urinary urgency, gastrointestinal distress, head-ache, and voice tremors.

Severe anxiety refers to a further narrowing of the perceptual field. The perceptual field of a severely anxious person becomes so greatly reduced that he or she can only focus on one particular detail or many scattered details. The individual has difficulty noticing what is going on around him or her even when it is pointed out. His or her behavior becomes more automatic and is aimed at finding relief. The individual often complains of severe somatic distress and may experience hyperventilation, a pounding heart, a sense of impending doom, and trembling.

Panic refers to a marked disturbance in anxiety in which the individual is not able to process what is happening within the self or the environment. The individual who panics may begin to scream and shout, become dazed or confused, begin running, or simply withdraw. His or her behavior is erratic, uncoordinated, and impulsive.

Relief of Anxiety

According to Peplau (1963), the energy of anxiety can be subsumed or discharged. She identified four major patterns of relieving anxiety. They are acting out, somatizing, "freezing to the spot," and learning. Acting out refers to observable behaviors that are displaced from one situation to another. They can include feelings and behaviors such as lashing out, crying, or laughing. These are behavioral indicators for "fight." Somatizing refers to the experi-ence of emotional conflict as physical symptoms such as headache, backache, or nausea. These are sensory indicators for helplessness. Freezing to the spot refers to the human response pattern of withdrawal and depression. With-drawal is the subsuming of energy from the environment into the self, whereas depression is the subsuming of aggression into the self. These are health indicators for "flight." Learning is a form of problem solving in which an individual talks out or figures out how to relieve feelings associated with threat or unmet expectations.

Crisis

There are two generic perspectives on crisis that are consistent with the Intersystem Model. They are the homeostatic and the developmental perspectives on the functioning of human systems (Shaw & Halliday, 1992).

Homeostatic models of crisis are driven largely by the principle of balance and negative feedback. Imbalance of the human system is presumed to be a serious risk factor for disintegration of the self. Human responses include those of ineffective coping and the experience of chaos. Aguilera (1990), Aguilera and Messick (1986), and Narayan and Joslin (1980) are nurses who typify a homeostatic perspective. They define crisis as a perceived or actual imbalance between perceived difficulty of a life challenge and an available and sufficient repertoire of coping skills.

Research from a homeostatic point of view shows that an individual in crisis is likely to report difficulties with self-care, interpersonal relationships, and deliberative action (Coler & Hafner, 1991; Geissler, 1984; Halpern, 1973; Leiter, 1992). In addition, Hoff (1989) described the person in a crisis as having a narrowed perceptual field in which he or she may be unable to think of events beyond a current problem and its resolution. The goal of action within the Intersystem Model is not simply the restoration of homeostasis. Rather, it is to assist the client to grow through an illness situation by expanding on his or her formerly "narrow" view of the situation.

Developmental models of crisis describe the principle of unfolding potential whereby crisis is presumed to be a decision point and an opportunity for growth. Rather than viewing imbalance of the human system as problematic, the developmental model interprets it as a challenge. Research on the human responses of persons from this perspective shows that an individual in crisis is likely to display creative learning as he or she incorporates change into his or her experience (Brown & Powell-Cope, 1991; Sohier, 1993; Steeves, Kahn, & Bennoliel, 1990). Within the framework of the Intersystem Model, the nurse can assist the client to make constructive decisions and try novel approaches to align (or realign) with the situational environment. The goal is the emergence of a new internal environment in the client—one that allows him or her to respond in new patterns that demonstrate an increased situational sense of coherence.

Adaptation

Kuhn (1974) defined adaptive behavior as behavior that changes the relation of a human system to its environment. This is done through

The detector (which receives information about the environment), the selector (which reflects the inner tendencies of the system to respond in one way rather than another), and the effector (which executes the behavior thus selected). (p. 42)

Viewed in this way, adaptive behavior results from a process similar to that described in the Roy Adaptation Model, in which adaptation is seen as a systematic series of actions directed toward some end (Roy & Andrews, 1991).

According to Lazarus and Folkman (1984), adaptation is a health outcome. For them, an adaptive outcome may be positive, negative, or neutral. A positive, negative, or neutral outcome in one domain of experience may or may not be reflected in another domain of experience. What may be positive, negative, or neutral in the short term may or may not be in the long run. In the Intersystem Model, adaptation is a state that is measured as the level of situational sense of coherence a person achieves with respect to a particular situation after intervention for a specific problem. Because this response is situation specific, it does not reflect an overall adaptation level as described in the Roy Adaptation Model, which may be more similar to the Level II concept *sense of coherence.*

Sullivan (1953) noted a linkage between anxiety and adaptation. He defined the self-system as an antianxiety system whose function is to prevent, reduce, or relieve anxiety. For him, a low level of anxiety represents an adaptive outcome in a threatening encounter.

Peplau (1968) believed that optimal, Level II anxiety could lead to a positive outcome following crisis. She and others believed that anxiety together with an appropriate learning environment creates opportunities for growth and healing. She proposed that nurses, as participants in the learning environment, could influence an individual's healing in the direction of growth.

Other researchers presuppose that increased energy and high subjective health signify a positive unfolding of health potential. They are likely to view this sign of positive health potential as evidence of the ongoing organization and reorganization of the biopsychosocial and spiritual domains of human experience within the niche of the individual.

Implications

Nurses need knowledge of anxiety, crisis, and adaptation because these human response trajectories underscore significant content for health and healing. That is, they represent a situational basis for adaptation. Nurses need to recognize and care for an individual's evolving human response trajectory in situations of health, illness, or death. Not to recognize and respond to the adaptive experience of an individual is to potentially undermine the individual's ongoing development of a situational sense of self-coherence.

Nurses bear responsibility for the diagnosis, treatment, and documentation of salient human response patterns that promote or interfere with the natural, learned unfolding of an individual's adaptive human responses. This responsibility is encoded in nurses' caregiver burden.

Clinical Example

The following clinical example describes an ongoing interaction between a patient with posttraumatic stress disorder and Donna Drugatz, RN, MSN, who at that time was a mental health clinical nurse specialist at the Visiting Nurse Association of Pomona at San Bernardino and assistant professor at Azusa Pacific University.[1] It illustrates how previous interactions become part of the patient's developmental environment as a new situational focus emerges.

Focus of Situational Interaction

Mrs. L was very fearful of having to face the offender in court. She had severe anxiety and said "I don't understand how he could hurt me so badly and just walk away."

Assessment

Developmental Environment

Biological Subsystem

Mrs. L is 44 years of age and is Caucasian. She was admitted to the Visiting Nurse Association (VNA) services after an auto accident. She had a cervical fracture with cervical fusion C-6 to T-1 treated with a bone graft from the left iliac crest and a halo brace. Three weeks later, her posterior incision was found to be red and edematous and a small opening was developing along the suture line. A sample for culture and sensitivity indicated a staphylococcus infection that was treated with IV antibiotics. She had allergic reactions to four of the antibiotics that were tried, which made her very upset and anxious about her physical status.

Psychosocial Subsystem

The patient was first evaluated for posttraumatic stress disorder by the psychiatric registered nurse 3 weeks after admission to VNA service. She gave the following account of the accident:

I was on my way to work and was approaching the entry ramp to the 210 freeway. The road conditions were slick because of rain. Traffic was slowing as cars were attempting the transition from ramp to freeway. I observed in my rearview mirror that the person in the car behind me was driving fast and erratic. Before he hit me I saw the male driver avoid hitting my car twice by driving off the ramp into the berm of the roadway. On his third approach, I braced myself before he rear-ended me and forced me off the roadway. He continued 100 yards down the ramp before pulling over. I don't know why, but as soon as I was hit, my hands went to my neck and I just held my neck. The man walked back to my car but his gait was very unstable. He said, "Move over and I will drive you to the nearest hospital. I can't afford to have the police find out about this. Just move over, I'm going back to get my wallet and then I'll take you." I realized that I needed help, but no one stopped in spite of the fact that I was constantly honking the horn. The man returned and said, "Move over." I said, "I can't move." He yelled "I said move over. I will take you somewhere to get help." A second man appeared on the scene. He asked if he could be of assistance. The man who hit my car said, "I'm going to drive her to get help." The second man said, "You are taking her nowhere. I have already telephoned for the police and paramedics." When he heard that, the first man ran toward his car, retrieved what appeared to be a leather briefcase, and ran off leaving his car. The second man who I consider to be my guardian angel stayed with me and calmed me until the paramedics arrived. Then he left and I never saw him again. I don't know his name. The police arrived and I was transported to the hospital.

When the psychiatric nurse evaluated Mrs. L, she found that she had recurrent, intrusive, and distressing recollections of the event with recurrent nightmares and anxiety and insomnia. She also had tachycardia, shortness of breath, nausea, dizziness, feeling of impending doom, tremulousness, increased startle response, and hypervigilance. She had psychological distress when riding in an automobile, characterized by panic attacks, decreased concentration, and the inability to "think straight" or solve problems because of her intense fear. She had continual tearfulness and was reliving the event day and night. Prior to the accident, the patient was very independent. She was divorced and had a son in the military and daughter in nursing school. She owned a condominium and was making monthly payments. She now sees herself as dependent on others for basic needs, such as bathing, toileting, and dressing, because of limited range of motion due to the halo brace and multiple reactions to medications. She says she hates taking anything more than an aspirin. Her brother had died 2 months before her accident of a myocardial infarction at the age of 40. Currently, her parents are with her to help her with her needs.

Spiritual Subsystem

Following her accident, Mrs. L's pastor had visited her frequently and, although he provided spiritual comfort, he recognized that she needed further assistance and suggested that she request mental health counseling.

Situational Environment

Intrasystem Analysis of the Client

Detector Function. Mrs. L said that on the day of the accident the offender had long hair, beard, and mustache and was greying. Despite the fact that he had changed his appearance for the identification photograph taken by the police, she had been able to identify him. He had shaved his face and had short, dark brown hair for the photograph. She was sure she would recognize him in court, however.

Selector Function. Although she was fearful of facing the offender in court, she knew it was necessary to have this man convicted because he already had two arrests for driving under the influence and had been drinking the morning of the accident. He had been driving his wife's car without a license and without insurance. He had also told the police that his friend had been driving the car and that he had been visiting his mother out of state the day of the accident. After Mrs. L had been able to identify the offender from the photo, the friend turned state's evidence to avoid perjury charges.

Effector Function. She had not been able to sleep the night before the hearing. She said that she had walked around the whole night while her mother and daughter massaged her legs, which were in spasm because she was so tense.

Intrasystem Analysis of the Nurse

Detector Function. The nurse had knowledge of the symptoms of posttraumatic stress disorder and had experience in working with patients with this problem. She was skilled in relaxation techniques and acupressure.

Selector Function. The nurse values her ability to assist patients at stressful periods. She is also concerned that justice is done.

Effector Function. The nurse accompanied the patient to the courthouse the day of the hearing.

Analysis

Identification of Stressors

Loss of physical health and independence
Recent loss of brother due to myocardial infarction
Experience of emotional trauma

Coping Resources

History of independent living
Supportive family
VNA services

Scoring on Situational Sense of Coherence

Comprehensibility = 3. Patient had good recall of the face of the offender and understood the symptoms of posttraumatic stress disorder.

Meaningfulness = 3. Patient felt it was important to bring the offender to justice despite the emotional trauma to her.

Manageability = 1. Patient had severe anxiety and was very fearful of the sight of the offender. She was unable to carry out her normal responsibilities.

Nursing Diagnosis

Posttraumatic stress disorder following an accident.

Nursing Goals—Client Goals

1. To decrease anxiety
2. To bring the offender to justice
3. To return patient to normal ability to handle life

Intersystem Interaction

Communication of Information

Mrs. L had already been instructed in relaxation techniques, and during the court appearance she needed reinforcement of them.

Negotiation of Values

Both patient and nurse had the same goals for the interaction and so no negotiation was needed.

Organization of Behaviors: Mutual Plan of Care

As soon as Mrs. L saw the offender, she went into a severe panic attack. The nurse talked her through the relaxation techniques she had mastered during the previous month—deep breathing, visualization, and focusing. Because of the severe muscle spasms the patient was experiencing, the nurse used acupressure as well as massage. The patient was able to identify the defendant as the man who caused the accident. In the month after the hearing, the patient and nurse continued to work on recall of the accident and relaxation techniques with gradual improvement of symptoms over time.

Evaluation of the Plan: Rescoring on SSOC

Comprehensibility = 3. No change.

Meaningfulness = 3. No change.

Manageability = 3. One month after the hearing, the patient returned to the hospital for resuturing of the incision line. She has called the agency to report full recovery of her physical strength and psychological well-being. She was able to return to work 7 months after the accident.

Note

1. Adapted with permission from Donna Drugatz.

References

Aguilera, D. C. (1990). *Crisis intervention: Theory and methodology* (6th ed.). St. Louis, MO: C. V. Mosby.

Aguilera, D. C., & Messick, J. (1986). *Crisis intervention: Theory and methodology* (5th ed.). St. Louis, MO: C. V. Mosby.

Angel, R., & Thoits, P. (1987). The impact of culture on the cognitive structure of illness. *Culture, Medicine, and Psychiatry, 11*, 465-494.

Apter, M. J. (1982). *The experience of motivation: The theory of psychological reversals.* London: Academic Press.

Apter, M. J. (1984). Reversal theory and personality: A review. *Journal of Research in Personality, 18*, 265-288.

Apter, M. J. (1991). A structural phenomenology of stress. In C. D. Spielberger, I. G. Sarason, J. Strelau, & J. M. T. Brebner (Eds.), *Stress and anxiety* (Vol. 13, pp. 13-22). New York: Hemisphere.

Arnold, M. B. (1960a). *Emotion and personality: Psychological aspects* (Vol. 1). New York: Columbia University Press.

Arnold, M. B. (1960b). *Emotion and personality: Neurological and physiological aspects* (Vol. 2). New York: Columbia University Press.

Arnold, M. B. (1970). Brain function in emotion: A phenomenological analysis. In P. Black (Ed.), *Physiological correlates of emotion.* New York: Academic Press.

Brown, M. A., & Powell-Cope, G. M. (1991). AIDS family caregiving: Transition through uncertainty. *Nursing Research, 40*(6), 338-345.

Buss, A. H., & Plomin, R. (1975). *A temperament theory of personality development.* New York: John Wiley.

Cantril, H. (1965). *Pattern of human concerns.* New Brunswick, NJ: Rutgers University Press.

Chess, S., & Thomas, A. (1984). *Origins and evolution of behavior disorders: From infancy to adult life.* New York: Brunner/Mazel.

Chess, S., & Thomas, A. (1986). *Temperament in clinical practice.* New York: Guilford.

Chess, S., & Thomas, A. (1990). The New York Longitudinal Study (NYLS): The young adult periods. *Child Psychiatry, 35*, 557-561.

Coler, M. S., & Hafner, L. P. (1991). An intercultural assessment of the type, intensity, and number of crisis precipitating factors in three cultures: United States, Brazil, and Taiwan. *International Journal of Nursing Studies, 28*, 23-235.

England, M. (1993). Implications of knowledge claims for continued study of perception. In J. Barnfather & B. Lyon (Eds.), *Stress and coping: State of the science and implications for nursing theory, research, and practice* (pp. 151-158). Indianapolis, IN: Sigma Theta Tau International.

England, M. (1996). Sense of relatedness and interpersonal network of adult offspring caregivers: Linkages with crisis, emotional arousal, and perceived health. *Archives of Psychiatric Nursing, 10*, 1-12.

Field, W. E. (1979). *The psychotherapy of Hildegard E. Peplau.* New Braunfels, TX: PSF Productions.

Geissler, E. (1984). Crisis: What it is and what it is not. *Advances in Nursing Science, 6*, 1-9.

Goldsmith, H. H., Buss, A. H., Plomin, R., Rothbart, M. E., Thomas, A., Chess, S., Hinde, R. A., & McCall, R. B. (1987). Roundtable: What is temperament? Four approaches. *Child Development, 58*, 505-529.

Halpern, H. (1973). Crisis therapy: A definitional study. *Community Mental Health Journal, 9*, 342-349.

Hoff, L. A. (1989). *People in crisis: Understanding and helping* (3rd ed.). Redwood City, CA: Addison-Wesley.

James, W. (1890). *Principles of psychology.* New York: Holt.

Kuhn, A. (1974). *The logic of social systems: A unified deductive system-based approach to social science.* San Francisco: Jossey-Bass.

Lazarus, R. S. (1966). *Psychological stress and the coping process.* New York: McGraw-Hill.

Lazarus, R. S., & Folkman, S. (1984). *Stress, appraisal, and coping.* New York: Springer.

Lazarus, R. S., & Launier, R. (1978). Stress-related transactions between person and environment. In L. A. Pervin & M. Lewis (Eds.), *Perspectives in interactional psychology.* New York: Plenum.

Leiter, M. P. (1992). Burn-out as a crisis of self-efficacy: Conceptual and practical implications. *Work & Stress, 6*(2), 107-115.

Loevinger, J. (1976). *Ego development.* San Francisco: Jossey-Bass.

Masters, W. H., & Johnson, V. E. (1966). *Human sexual response.* Boston: Little, Brown.

Maturana, H. R., & Varela, F. J. (1987). *The tree of knowledge: The biological roots of human understanding.* Boston: New Science Library.

May, R. (1950). *The meaning of anxiety.* New York: Ronald Press.

McMillan, D. W., & Chavis, D. M. (1986). Sense of community: A definition and theory. *Journal of Community Psychology, 14,* 6-23.

Mead, G. H. (1934). In C. W. Morris (Ed.), *Mind, self, and society: From the standpoint of a social behaviorist.* Chicago: University of Chicago Press.

Narayan, S., & Joslin, D. (1980). Crisis theory and intervention: A critique of the medical model and proposal of a holistic nursing model. *Advances in Nursing Science, 2,* 27-39.

Nichols, M. P., & Zax, M. (1977). *Catharsis in psychotherapy.* New York: Gardiner.

Peplau, H. E. (1963). Working definition of anxiety. In S. Burd & M. Marshall (Eds.), *Some clinical approaches to psychiatric nursing.* New York: Macmillan.

Peplau, H. E. (1968). Psychotherapeutic strategies. *Perspectives on Psychiatric Care, 6,* 264-278.

Rothbart, M. K. (1984). Longitudinal home observation of infant temperament. *Developmental Psychology, 22,* 356-365.

Roy, C., & Andrews, H. A. (1991). *The Roy adaptation model: The definitive statement.* Norwalk, CT: Appleton & Lange.

Scheff, T. J. (1984a). The taboo on the coarse emotions. *Review of Personality and Social Psychology, 5,* 1946-1969.

Scheff, T. J. (1984b). The theory of catharsis. *Journal of Research on Personality, 18,* 238-264.

Scheff, T. J. (1985). Universal expressive needs: A critique and a theory. *Symbolic Interaction, 8,* 241-262.

Schulz, R., Heckhausen, J., & Locher, J. (1991). Adult development, control, and adaptive functioning. *Journal of Social Issues, 47,* 177-196.

Shaw, M. C., & Halliday, P. H. (1992). The family, crisis, and chronic illness: An evolutionary model. *Journal of Advanced Nursing, 17,* 537-543.

Shotter, J. (1986). A sense of place: Vico and the social production of social identities. *British Journal of Social Psychology, 24,* 199-211.

Sohier, R. (1993). Filial reconstruction: A theory of development through adversity. *Qualitative Health Research, 3,* 465-492.

Steeves, R. H., Kahn, D. L., & Bennoliel, J. Q. (1990). Nurses' interpretation of the suffering of their patients. *Western Journal of Nursing Research, 12,* 715-731.

Strauss, G. (1956). *The social psychology of George H. Meade.* Chicago: University of Chicago Press.

Sullivan, H. S. (1953). *The interpersonal theory of psychiatry* (2nd ed.). New York: Norton.

Thomas, A., & Chess, S. (1977). *Temperament and development.* New York: Brunner/Mazel.

Thomas, A., Chess, S., & Birch, H. G. (1968). *Temperament and behavior disorders in children.* New York: New York University Press.

Van der Velde, C. D. (1985). Body images of one's self and of others: Developmental and clinical significance. *American Journal of Psychiatry, 142,* 527-537.

7

Theoretical Foundations for the Spiritual Subsystem

Julia D. Emblen

Differentiation of Terms: *Spiritual* **and** *Religious*
Systems Analysis of the Spiritual Subsystem
Clinical Example

The spiritual subsystem lies at the heart of the description of person in the Intersystem Model. An understanding of what this subsystem consists of and how it interrelates with the other body subsystems is the focus of this chapter. A systems analysis is used to illustrate how life events are processed through the spiritual subsystem and produce either spiritual well-being or spiritual distress. In the Intersystem Model, spiritual is used as the descriptor for the inner center of the person. Religious practices will flow out of that center in expressions of the internal spiritual system.

This spiritual center of the person is also what gives answers to the questions of "Who am I? Where am I going? What's it all about?" (Olthuis, 1989). Within this center of the person are fundamental beliefs that form the basis for one's worldview. These beliefs have developed in interaction with others within a particular culture and serve as a channel through which one's beliefs provide direction and the meaning of life. It is the basis for one's fundamental decision-making process. One's response to life situations is formed by these internally held beliefs. Thus, as one considers interactions with another person, an understanding of how these core beliefs are developed is important. They will form who the person is at a fundamental level and will direct one's responses to life situations.

Differentiation of Terms: *Spiritual* and *Religious*

Because there has been some overlap in the past between the use of the terms *spiritual* and *religious,* it is important to define how the term spiritual is used in nursing language today. A variety of transcendent values can be included as spiritual, either with or without inclusion of a religious emphasis referring to deity or supreme being with related beliefs and practices. Emblen (1992) has done an analysis of the use of these terms showing how separation between them has occurred during the past three decades. In the 1960s, the terms religious and spiritual were used synonymously. Today, the nursing literature indicates that the umbrella term is spiritual. This term encompasses the ideas of a "personal life principle which animates transcendent quality of relationship with God or god being" (Emblen, 1992). Others have defined spirituality as the "propensity to make meaning through a sense of relatedness to dimensions that transcend the self in such a way that empowers and does not devalue the individual" (Reed, 1992, p. 350). The term religious, on the other hand, is defined as engaging in a "system of organized beliefs and worship that the person practices" (Emblen, 1992). In current practice, the umbrella term is spiritual with the term religious designated as one component of it.

Several societal shifts have contributed to the separation of meaning in these terms. Cultural pluralism has contributed to the expansion and broadening of the understanding of the spiritual with the addition of Eastern religious thought and other views of deity. Secular humanistic values have also been instrumental in the shift from a religious focus. The humanist acknowledges a need to have a belief in something toward which one can devote one's energy but focuses on self-knowledge and self-actualization rather than on a supreme being (Maslow, 1970). Self-knowledge, or the "knowing and understanding of one's self and the universe in which one exists, knowing what one is, where one comes from, where one is going, and what purpose one is to serve" (Edgar, 1989, p. 17), is vital so that self-actualization can be pursued. This perspective has a diminished focus on the vertical aspect of the spiritual dimension because this self-exploration is not transcendent of self.

This broadened use of the term spiritual is essential for accurately representing the components currently included within the spiritual subsystem. Previous classifications of behaviors were designated as religious and psychosocial (Emblen, Fitchett, Farran, & Burck, 1991) or religious and existential (Ellison, 1983). Such arbitrary combinations lack the degree of coherence that is necessary for further development. Using the term spiritual within a broader context aids in specifying ideas and thoughts consistently used in relation to spiritual as a singular concept rather than in conceptual combination with terms borrowed from other disciplines. In analyzing the spiritual concept, Mansen (1993) has identified the need to distinguish between the spiritual dimension and the psychosocial dimension, because without a spiritual concept the nurse can too easily deny that persons have spiritual needs that are distinctly different from their psychosocial needs. A research study was designed to investigate this distinction and is reported in Chapter 16.

Systems Analysis of the Spiritual Subsystem

A systems analysis of the functioning of the spiritual subsystem is shown in Figure 7.1. The inputs into the spiritual subsystem are life events and life changes that occur to the person. These environmental inputs are filtered through the body's other subsystems, the biological and psychosocial, to reach the inner person, the spiritual subsystem. Life events that can serve as input to the spiritual subsystem include suffering, such as loss or illness and alterations in lifestyle. The events are processed through the person's beliefs and values. In systems language, these beliefs and values are viewed as the throughput and are expressed as measurable outputs such as a person's response to deity and various religious rituals and practices. The processing of life events can result in changes in a person's equilibrium within the spiritual subsystem. The person's view of deity may undergo serious question. Also, the person's religious rituals and practices may be altered. These alterations can cause significant changes in the person's situational sense of coherence (SSOC). Positive adaptation within the spiritual subsystem will lead to spiritual well-being; negative adaptation within the subsystem can result in spiritual incoherence and produce spiritual distress.

Each of the processes is expanded in this chapter to illustrate how typical interactions occur. Altered coherence with respect to religious practices and affective responses is identified. A clinical example illustrating the role of the nurse in detecting and preventing spiritual distress is included.

Input

Life Events

Life events, both positive and negative, are the pivotal inputs to the spiritual subsystem. These reach the spiritual subsystem through the other body subsystems, the biological and psychosocial. Changes in the biological subsystem—physiological changes—that result in disease processes are significant life events that can cause disequilibrium in the spiritual subsystem. Other life events are altered through the psychosocial subsystem. One group of life events is developmental. For example, the birth of an infant is a life event to both the infant and the parents. The beginning of school is another life event for the child that may be either positive or negative, depending on the perspective of the child and the parents. Adolescents experience a number of events that affect their lives—for example, developing independence and new experiences of relationships with the opposite sex as well as with parents. Events for young adults typically include marriage and childbearing. Life events for the middle adult may include positive aspects such as marriage of children as well as negative events such as divorce or death of spouse. For the older adult, the major life event is retirement, which may be viewed negatively or positively. Moves, job changes, and financial gains and losses are other types of life events that occur throughout the life span of individuals and family groups.

Life Changes

Although some life events occur in a timely, developmental fashion with expected or anticipated changes in life experiences, other life events may be sudden and unexpected and precipitate major crises. Illness and its accompanying suffering sometimes precipitate major crises in the experiences of individuals and families. Losses such as immobility owing to accidents or other causes such as strokes may precipitate personal or family crisis. Illnesses that lead to death, such as heart attacks or cancer, may also lead to crises. Sometimes even normal developmental life events precipitate crises. For some young adults, the responsibility of parenthood may produce a psychological crisis.

Life events producing significant change in any body system and at any time in life may precipitate a maladaptive response. Personal religious beliefs and values may be stabilizing for some individuals at this time. For others, these life events may precipitate a crisis in the spiritual subsystem. These beliefs and values are designated as the *throughput* in this systems model and are discussed in the following section.

Throughput

In Figure 7.1, the throughput is depicted as the spiritual belief system and the values that each person holds that produce specific religious practices. These are affected by the life events described in the preceding section. A

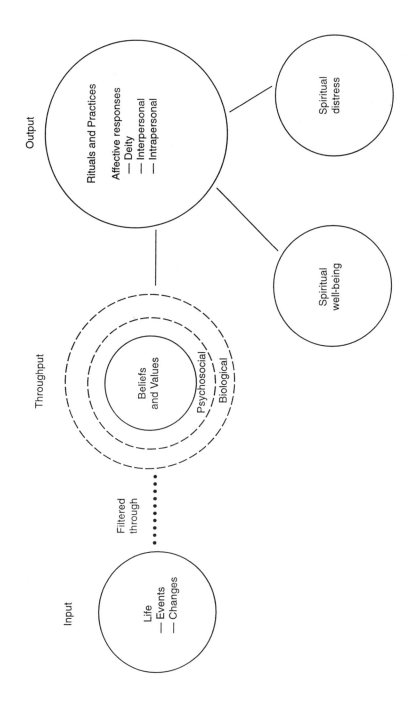

Figure 7.1. Model of Spiritual Subsystem

111

person's cultural orientation will have a significant effect on forming these beliefs and values, and the outcomes can be quite different. This chapter examines the belief systems of three selected religions—Buddhism, Christianity, and Judaism—to explore these differences.

Belief Systems

Buddhism. Buddhist beliefs are recorded in the Theravada or Pali canon, which includes the teachings of the Buddha as well as those of adherents to Buddhism. Key beliefs (Braden, 1954) include the Four Noble Truths:

1. The Noble Truth of suffering
2. The Noble Truth of the Cause of suffering
3. The Noble Truth of the Cessation of suffering
4. The Noble Truth of the Path that leads to the cessation of suffering.

The other important teachings describe the Holy Eightfold Path:

1. Right Views
2. Right Aspirations
3. Right Speech
4. Right Conduct
5. Right Livelihood
6. Right Endeavor
7. Right Mindfulness
8. Right Meditation

Christianity. Christian beliefs are set forth in the New Testament of the Bible. It contains both the words of Christ, the central figure, and teachings set forth by Christ's apostles or close followers. The Apostle's Creed evolved during the late sixth and early seventh centuries when Christians were asked to succinctly state what they believed regarding Christianity (Runia, 1963). It is based on the Nicene Creed (381 C.E.), one of the earliest recorded creedal statements of belief. The Apostle's Creed, often repeated in Christian worship services even today, is one of the most succinct statements of beliefs common to many adherents of the Christian religion. It follows:

I believe in God, the Father Almighty, Maker of heaven and earth.

And in Jesus Christ, His only Son, our Lord, who was conceived by the Holy Spirit, born of the Virgin Mary, suffered under Pontius Pilate, was crucified, died, and was buried. He descended into hell. The third day He arose again from the dead. He ascended into heaven and sits at the right hand of God, the Father Almighty. From thence He will come to judge the living and the dead.

I believe in the Holy Spirit, the holy catholic church, the communion of saints, the forgiveness of sins, the resurrection of the body, and the life everlasting (*The Westminster Confession of Faith,* 1643-1648).

Judaism. Judaism focuses on the belief that the Jewish people have an important role as God's chosen people in redemption for the world, with the introduction of the Messiah coming from the Jewish lineage. Jews' basic beliefs come from the Torah, the first five books of the Bible, and the Talmud, an encyclopedic commentary on the Torah and an authoritative guide to Jewish tradition. They believe in the Ten Commandments as God's word given to Moses on tablets of stone while he was talking with God on Mount Sinai (Cohen & Mendes-Flohr, 1987). They are as follows:

1. You shall have no other gods before me.
2. You shall not make for yourself an idol in the form of anything in heaven above or on the earth beneath or in the waters below. You shall not bow down to them or worship them; for I, the Lord your God, am a jealous God, punishing the children for the sin of the fathers to the third and fourth generation of those who hate me, but showing love to thousands who love me and keep my commandments.
3. You shall not misuse the name of the Lord your God, for the Lord will not hold anyone guiltless who misuses His name.
4. Remember the Sabbath day by keeping it holy. Six days you shall labor and do all your work, but the seventh day is a Sabbath to the Lord your God. On it you shall not do any work, neither you, nor your son or daughter, nor your manservant or maidservant, nor your animals, nor the alien within your gates. For in six days the Lord made the heavens and the earth, the sea, and all that is in them, but he rested on the seventh day. Therefore the Lord blessed the Sabbath day and made it holy.
5. Honor your father and your mother, so that you may live long in the land the Lord your God is giving you.
6. You shall not murder.
7. You shall not commit adultery.
8. You shall not steal.
9. You shall not give false testimony against your neighbor.
10. You shall not covet your neighbor's house. You shall not covet your neighbor's wife, or his manservant or maidservant, his ox or donkey, or anything that belongs to your neighbor (New York Bible Society International, 1978).

Values

Personal values are another throughput identified in Figure 7.1. Four general considerations that determine or affect values are identified—Eastern-Western, transcendence, relationship, and developmental stages.

Eastern-Western. The Eastern-Western views of the spiritual dimension have been distinguished by Farran, Fitchett, Quiring-Emblen, and Burck (1989). The Eastern view of spirituality is that the spiritual reflects the totality of the human being and becomes an overarching umbrella unifying all the parts of the human being. If the human being is conceptualized as composed of biological, psychological, sociological, and cultural components, the spiritual component would surround all, providing a holistic harmonization. Figure 7.2 depicts the relationship of the spiritual subsystem with the other subsystems composing the person in both Eastern and Western thought.

In the Western view, the spiritual dimension is considered as one system among a group of systems. In this view, the spiritual would be one part as are the biological, psychological, sociological, and cultural components. All component systems operate on the same level and are thus integrated within the human being.

The difference in the valuing of the spiritual is apparent when making choices regarding every aspect of life. For those who hold that the spiritual is an integrated part, spiritual considerations would be considered along with all the other components when making choices regarding life and care. For those who hold that the spiritual is the overarching dimension that unifies all the others, spiritual values would be used to establish priorities in all the other areas as they interface with the other human components under consideration.

The relationship of the spiritual subsystem to the other subsystems in the Intersystem Model is similar to the view presented as part of the Western culture in that there are distinct subsystems. In the Intersystem Model, however, the relationship of the spiritual subsystem to the others is different. The spiritual subsystem as conceptualized by Stallwood and Stoll (1975) is seen to be the very core of the person's being (see Figure 7.2). From this position, it forms the center of who the person is and has a pervading influence on the other subsystems. Thus, it has more than just an effect on the psychosocial and biological subsystems; it forms the very core of who the person is. Any assault on this core will have a profound effect on the whole person. At the same time, the spiritual subsystem is in constant communication with the other subsystems and is constantly receiving input from them. For these reasons, Carson (1989) states that the perspective that one's spirituality is at the core of one's being carries with it the sense of the universality of the spiritual dimension as a concern for nursing.

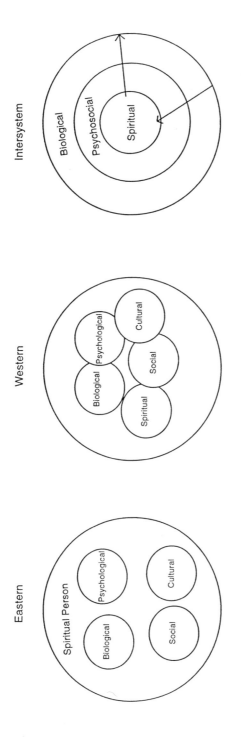

Figure 7.2. Options for Viewing the Spiritual in Eastern, Western, and Intersystem Models

Transcendence. Because the spiritual dimension is abstract, it follows that related values will similarly be abstract. The word *transcend,* according to *Webster's Third New International Dictionary* (Grove, 1986), indicates rising above or going beyond the limits or exceeding. As the term is used in the context of the spiritual, it typically refers to metaphysical aspects of Aristotelian or Kantian thought. Perhaps the clearest comparison in the context in which it is used here is between *transport* and *transcend.* In transport, a person is literally moved, whereas in transcendence, the person may physically remain stationary while his mind moves upwards or beyond a given moment of experience.

Emblen and Halstead (1993) interviewed patients, nurses, and chaplains to identify what they designated as a spiritual need. Based on responses to the question, "What would you describe as a spiritual need?," patients identified a "need more powerful than earthly being" and "something to hold." Nurses identified a "dimension other than physical-mental" and chaplains identified a "sense of direction," "assurance of God's presence," "contact with God," and "inner need-inner person." On the basis of the context from which these phrases were taken, the category of transcendence was designated. Transcendence is an aspect of valuing that is essential when considering abstract values such as love, joy, pain, and sorrow. While experiencing pain on the biological level, the human being may be able to spiritually rise above the pain of the moment to experience love or joy on a higher level of spiritual experience. Thus, the transcendence aspect can literally change a negative experience to a positive valuing of the experience.

Relationship. Another aspect related to spiritual value is relationship. This term may be expressed in religious terms such as *fellowship* or *communion.* In a survey of a group of surgical patients, it was found that they identified the relationship of the nurse as presence (Emblen & Halstead, 1993). In the same survey, the patients identified the chaplain's relationship as the need for a friend, being around after surgery, and providing support by way of consolation, encouragement, and comfort.

Relationship affords a connectedness to other human beings either in a biological familial context or in a personal one-on-one or small group context. The relationship signifies the fact that people care. It also provides the sense that the person is not abandoned during points of crisis and illness.

Developmental Stages. According to some theorists, there are moral developmental stages that contribute to the values persons hold at given points in their personal development. According to Kohlberg (1968), the individual moves through six possible moral stages: (a) heteronomous morality; (b) individualism, instrumental purpose, and exchange; (c) mutual interpersonal expectations, relationships, and interpersonal conformity; (d) social system and conscience; (e) social contract or utility and individual rights; and (f) universal ethical principles. Each person's moral development, however, progresses at different rates, with many never reaching the sixth or highest stage.

Fowler (1981), another theorist, has identified seven phases of spiritual development: (0) undifferentiated, (1) intuitive-projective, (2) mythic-literal, (3) synthetic-conventional, (4) individuative-reflective, (5) conjunctive, and (6) universalizing. These phases are described in Chapter 3 in relation to how they are learned within a particular culture.

Values are developed and modified throughout the life span, probably in accord with individual personal characteristics as well as personal life experiences, events, and related values. Both the life events and experiences as well as their interpretation are affected by the developmental stage of the individual's spiritual values.

Output

Output is considered in terms of rituals and practice and affective responses.

Religious Practices

Religious practices are outputs of one's religious beliefs and values (see Figure 7.1). They arise directly from one's belief systems. Two of the most commonly identified religious practices in Western Christianity are prayer and religious readings (Emblen & Halstead, 1993). Others include group or private worship, participation in sacraments such as communion and confession, singing, meditation, fasting, and giving and offering that entail a type of personal sacrifice. All religions have specific rituals and practices arising from their belief system. Figure 7.3 indicates selected practices common to Buddhism, Christianity, and Judaism. Other authors have identified some of the most specific health-related religious practices by religious group (Ellis & Nowlis, 1985; Murray & Zentner, 1985; Pumphrey, 1977).

Table 7.1 presents a general listing of religious practices and rites that are often practiced by members of religious groups. Not all are practiced by any one group, but the listing represents a cross section of those typical of Eastern and Western religious group practices.

Affective Responses

Figure 7.3 identifies three kinds of affective responses: (a) to deity, (b) interpersonal, and (c) intrapersonal. The spiritual subsystem has the potential for transforming responses in all three levels. The response to deity or any

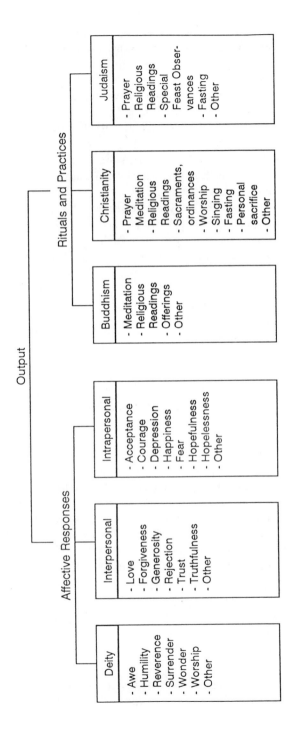

Figure 7.3. Output Identified by Affective Responses and Rituals and Practices

Table 7.1   Examples of Religious Practices

Prayer
Meditation
Worship—special days may entail activity modification (e.g., rest from usual work)
Religious reading
Teaching or preaching
Music-praise/adoration, and so on
Sacraments
Fasting or feasting—omit or add selected items (e.g., meat)
Personal sacrifice—financial, time, or self
Rites
   Birth
      Cleansing—baptism
      Naming
      Presentation
      Circumcision
   Puberty—confirmation
   Marriage—sanctifying relationship
   Sickness
      Use of special oil and substances for anointing
      Touch
   Death
      Cleansing
      Burial or cremation
      Mourning (time, activity, and dress)
Use of icons, pictures, and symbols to represent significant events, people, or experiences
Use or no use of substances including medicines, alcohol, tobacco, coffee, tea, and various
   other therapies
Wearing special clothes—(e.g., head coverings or drapes)
Others—variations with religious beliefs with respect to specific religious expectations and
   prohibitions

godlike being typically elicits responses that relate to mystical experiences. Maslow (1970) studied peak religious experiences to identify common characteristics. He identified an emotional response to the holy or numinous (deity or spirit) as awe, humility, reverence, surrender, wonder, or worship (see Figure 7.3). Maslow observed that, during peak religious experiences, individuals lose fear and anxiety and become more loving, accepting, and honest. These affective responses demonstrate the linkage between the response to deity and intrapersonal and interpersonal responses. Because the person responds as a unity, the response to deity is related to these factors and is filtered through the biological and the psychosocial subsystems. The personal characteristics of the individual, such as general sensitivity, hardiness (Lambert & Lambert, 1987), and right- and left-brain orientation (Springer &

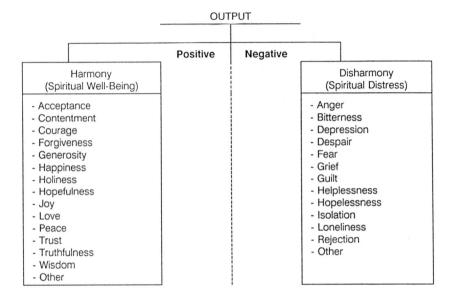

Figure 7.4. Output Identified by Harmony and Disharmony

Deutsch, 1981), as well as physical vitality, influence affective or feeling responses.

With a broader definition of spiritual focusing on transcendent emotions arising from religious beliefs and values, many additional spiritual responses need to be identified. Emotional responses arising from the spiritual context may be classified as directly religious emotions when they refer to God or identified religious rituals associated with specific values such as sin or guilt. These can be intrapersonal responses such as hopefulness or interpersonal responses such as love and anxiety (Emblen et al., 1991). Figure 7.3 identifies selected affective responses to deity as well as intra- and interpersonal responses. Those that are more related to others, such as love, forgiveness, generosity, rejection, trust, and truthfulness, are designated as interpersonal. Those that are more related to one's internal being, such as acceptance, courage, depression, happiness, fear, hopefulness, and hopelessness, are designated intrapersonal.

Harmony of the Spiritual Subsystem. Figure 7.4 suggests that there are harmonious positive spiritual emotions, such as acceptance, contentment, courage, faith, forgiveness, generosity, happiness, holiness, hopefulness, joy, love, peace, trust, truthfulness, wisdom, and so on, and negative spiritual emotions, such as anger, bitterness, depression, despair, fear, grief, guilt, helplessness, hopelessness, isolation, loneliness, rejection, and so on. Personal affective responses to life events are the product of personal characteristics filtered through a person's beliefs and values. The affective responses of individuals, along with their religious rituals and practices (if any are observed) produce either spiritual well-being or spiritual distress. In real persons this is not an either-or condition. Rather, persons experience greater degrees of well-being or distress depending on previous input and processing of the system throughput.

For example, when a young man suddenly loses a leg as the result of a car accident, his beliefs and values are challenged. If, when confronted by his own finiteness, he turns to a childhood belief in God as a loving shepherd, his personal sensitivity produces an affective reverence and surrender to the continuing care of the shepherd. Such a response to deity is likely to elicit positive emotional responses, such as acceptance, contentment, and courage to cope with the change and adjust to using a prosthetic limb.

A different scenario might develop for the same situation. If another young man sustained the same injury but experienced no belief in deity, he might attribute the accident to fate. He then would respond through his psychosocial subsystem. He might respond angrily, particularly if he were normally a volatile person. Other negative emotions might follow, such as bitterness and depression, and eventually the young man might commit suicide. These situations have been deliberately designed to reflect the polar emotional responses. Typically, personal responses would include some positive and some negative emotions, varying with individual characteristics and values as the experiences of coping and adaptation fluctuate from day to day.

A person's affective responses typically are consistent with his or her religious beliefs and values. Each new life event, change, or crisis poses a threat to the stability of the spiritual subsystem. When the affective responses differ from the expected in light of the personal presence or absence of particular religious beliefs and values, a degree of incoherence occurs. If the inconsistency continues, spiritual distress may result. It is important for a nurse to assess a person's spiritual subsystem carefully to determine whether spiritual well-being or distress is present.

Spiritual Well-Being. Moberg (1984) became interested in the sociological implications of spiritual responses and developed social indicators of what was termed *spiritual well-being.* For him, a social indicator facilitates concise judgments about a major aspect of society. He found that, in quality of life studies, the religious aspects were missing. His research on these issues identified the following components as implicit to spiritual well-being:

Mental-emotional states

Feelings

Relationships and reality orientation

Meaning, destiny, goal attainment

Autonomy

Religious faith

Ellison and Smith (1981) expanded the vertical and horizontal components of Moberg's (1984) research on spiritual well-being. They compared the measurement of spiritual well-being to the use of a stethoscope, which measures the function of the heart. They indicated that spiritual well-being reflects spiritual health but does not measure the spirit itself. They then developed a 20-item questionnaire composed of religious and existential items of well-being. An example of a religious item is "I don't find much satisfaction in private prayer with God." An existential item is "I don't know who I am, where I came from, or where I'm going."

Such researchers have focused on the need to explore spiritual wellness and illness behaviors. The abstract nature of the spirit, however, makes concise measurement of wellness or illness practices difficult. It is apparent, however, that there is a great need for measuring some of the reflections of the spirit because so much of human behavior includes the spiritual dimension as either a major directional force or an integrational component (depending on one's Western or Eastern view). Further work in this area is needed to provide means to measure the effect of harmony within the spiritual subsystem on personal behaviors.

Spiritual Distress. Nurses have identified spiritual distress as an important parameter for assessment. Human behaviors have been assessed and itemized within the nursing diagnosis of spiritual distress. According to the North American Nursing Diagnoses Association, spiritual distress is defined as a "disruption in the life principle which pervades a person's entire being and which integrates and transcends one's biological and psychosocial nature" (Kim, McFarland, & McLane, 1987, p. 55). A major contributing factor responsible for spiritual distress has been identified as a "challenged belief and value system." Based on the systems theory presented in this chapter, it is easy to see how a challenge of illness might produce negative affective responses such as despair, loneliness, helplessness, and grief. Without a religious belief system with accompanying religious rituals and practices that focus on acceptance of life events as a surrender to deity, the common human responses of anger might readily surface. Even with a religious value system with regular sacramental practices, negative emotions quickly take over. Instead of positive emotions of happiness, joy, and peace, the opposite response of anger, depression, and despair may become pervasive.

Using a systems analysis, the components of the spiritual subsystem have been designated as input—life events and changes; throughput—beliefs and values; and output—rituals and practices and affective responses to deity. Life cycle changes that cohere can produce spiritual well-being; if harmony is not found within the spiritual subsystem, incoherence can result and produce spiritual distress. Affective responses resulting from the harmony or disharmony of the spiritual subsystem can result in positive or negative emotions. Coherence of religious beliefs and values, in light of life events and crisis change, is related either to spiritual well-being or to spiritual distress and is reflected in a person's SSOC.

Clinical Example

The following example is excerpted from a case study written by Susan K. Fuentes,[1] a perinatal grief counselor and graduate student nurse at Azusa Pacific University. It illustrates the role of the nurse in detecting and preventing spiritual distress.

Focus of Situational Interaction

The perinatal nursing supervisor (perinatal grief counselor) has been called to see a patient who had a cesarean section delivery of her 9-month gestation stillborn son. She is in the Post Anesthesia Care Unit with her husband at the bedside. The patient is agitated and starts to cry, asking, "Where's my baby?" The perinatal grief counselor (nurse) approaches Mrs. S and says, "I am very sorry about your baby. I am here to help you decide what you want done for the baby. There are arrangements that need to be made." When the patient appears agitated and begins crying, the nurse initially thinks the patient is exhibiting pain but then realizes that there is more to the patient's behavior than physical pain.

Assessment

Developmental Environment

Biological Subsystem

Mrs. S is experiencing postoperative pain and nausea and she is frightened because she cannot move her legs. She is crying intermittently.

Psychosocial Subsystem

Mr. and Mrs. S have been married for only 6 months. They were married primarily because of the pregnancy, although they have expressed love for each other. They are both in their early twenties and are working professionals. Mrs. S has parents who live in the area and are on the way to the hospital.

Spiritual Subsystem

Mr. and Mrs. S were both raised in the Catholic tradition but are not currently attending any church. Mrs. S says that they are Catholic, but that they have not been practicing. Mrs. S says that her parents have strong opinions about their Catholic beliefs and traditions and that they have a strong influence on her.

Situational Environment

Intrasystem Analysis of the Patient

Detector Function. Mrs. S does not know about the postsurgical and postpartum recovery process or about postmortem procedures. She has not experienced a loss such as this before and does not know what to expect of herself. She is unsure if she should see the baby.

Selector Function. Mrs. S values Catholic beliefs and wants her baby to be baptized. She feels her baby is in heaven. She is willing to use available resources and has respect for authority figures. Her family is very important to her. She expresses love for her husband but fears that their marital relationship is weak and fragile.

Effector Function. She is alert and oriented but is acting very anxious. She is seeking information about the condition of her baby. Her communication with her husband is hesitant.

Intrasystem Analysis of the
Perinatal Grief Counselor (Nurse)

Detector Function. The nurse has a good knowledge base of obstetrics and post-cesarean-section care and is acquainted with the grief process and perinatal losses. She is a trained perinatal grief counselor and has good listening skills.

Selector Function. She values the marital relationship and believes communication to be the key to that relationship. She values the grief process as an important progression toward healing. She values the patient's belief system and recognizes the support it gives to the patient. She has a Protestant Christian belief system that views premarital sex as a sin. Nevertheless, she values her efforts to care for patients in a nonjudgmental way.

Effector Function. The nurse is able to assess and document the grief process. She is able to offer prayer or other spiritual support as they seem appropriate because she is confident in her own spirituality. She can teach the patient the signs and symptoms of grief and grief stages. She is authorized to perform baptisms.

Analysis

Identification of Stressors

Patient has just learned that her baby was stillborn.
Patient is experiencing pain and nausea and is anxious.
The marital relationship is weak and communication is poor.
Patient is not practicing her religious faith.

Coping Resources

Couple expresses love for each other.
Patient's family is supportive.
Patient is willing to avail herself of spiritual help.
Patient has a strong attachment to the baby.

Scoring on Situational Sense of Coherence

Comprehensibility = 1. Mrs. S has little knowledge about recovery from a cesarean section or from grief. She does not understand postmortem procedures for her baby. She does not understand why she has pain and cannot move. She is expressing confusion about her own and her husband's feelings of guilt, disbelief, and sadness.

Meaningfulness = 2. Mrs. S is eager to find answers to her questions and is seeking people to give her information. She is bonded to her baby to some degree and wishes to see him, although she is also afraid of this. The husband appears not to want to learn or invest himself in this experience. He sees this as "something horrible I just want to forget." He has also expressed fear of making himself vulnerable and communicating how he feels to his wife.

Manageability = 1. Although Mrs. S has some coping resources within herself, she does not realize this or trust her own ability to meet this demand. She is looking to her husband for support and not getting it. The parents may provide some support within their own belief system.

Nursing Diagnoses

(a) Shock and fear of the unknown secondary to a grief reaction related to the loss of her baby; (b) alteration in family process related to impaired verbal communication about the loss of the baby.

Nursing Goals

1. To encourage parents to view infant
2. To assist parents in grief process

Intersystem Interaction

Communication of Information

The perinatal grief counselor (nurse) explained to Mr. and Mrs. S the postmortem procedures of measuring, weighing, footprinting, taking pictures, autopsy consent, option of baptism, chaplain visit, funeral services, holding infant, and further grief counseling and encouraged their active participation. She described the infant's appearance and offered to bring the infant when they were ready. She instructed them in the stages of grief and what to expect as they go through the grieving process.

Negotiation of Values

Although the parents were reluctant to see the baby, the perinatal grief counselor (nurse) explained the advantages of seeing their baby as related to healthier grieving and easier recovery. She encouraged them to see their infant and identified herself as someone who was accessible to assist them in the viewing. Although the parents were not ready to view the baby, the nurse recognized that infant baptism is very important to Catholics and asked if they would like to have the baby baptized. Mrs. S said,

"Could you? I was wondering how we could make sure he got baptized. It is very important in our religion, and my parents will be upset if he isn't baptized." Mr. S said, "It would be great if you could." The nurse helped the parents accept their feelings as normal and to accept the need for communicating feelings with each other and with the grief counselor.

Organization of Behaviors: Mutual Plan of Care

The patient was asked what she wanted in relation to this experience. She identified the following things that she wanted to accomplish. From these expressed wants, the patient and nurse negotiated to develop the following plan:

1. Mrs. S will verbalize understanding of the postmortem procedure as well as recognize and carry out her options such as having a priest visit, having a room on another floor, having her infant baptized, postponing the funeral until her recovery, and so on.
2. Mrs. S will see and hold her infant. She will ask for her baby's mementos and pictures before going home.
3. Mrs. S will become more comfortable in verbalizing her feelings and in understanding the stages of grief.
4. Mrs. S will give support to her husband and seek support from him. If this support is not available, she will find other caring resources in the hospital or community.

In preparing to baptize the baby, the perinatal grief counselor (nurse) said, "When I do the baptism, I like to use the birth name. Have you named him yet?" Mr. S answered, "Yes, his name is Christopher Bryan. Thank you so much." The nurse asked if they wanted to witness the baptism, but because they were not ready to view the baby, the nurse baptized the baby and brought a certificate of baptism to the parents. A priest was located through the hospital chaplain, and he came to visit and planned to follow up with the couple after release from the hospital. The parents viewed the baby and made decisions regarding him.

Evaluation of the Plan: Rescoring on SSOC

Comprehensibility = 3. Mrs. S has accepted the instruction and taken responsibility for decisions relating to herself and the infant. She verbalized understanding of the grief process and accepts her feelings as normal and necessary.

Meaningfulness = 2. Although Mrs. S is very invested in this experience, the husband remains reluctant to become involved. He has become more interested in what is going on and has contributed to some of the decisions but is still wary of communicating or accepting his own feelings of grief.

Manageability = 3. Mrs. S developed good rapport with the perinatal grief counselor and used her appropriately as a resource. The baptism of her infant was very significant and a great comfort to her. The visit of the priest was seen as supportive, and he gained the husband's trust so that he views him as a source of support in the future.

Note

1. Adapted with permission from Susan K. Fuentes.

References

Braden, C. S. (1954). Buddhism. In *The world's religions* (pp. 118-134). New York: Abingdon.

Carson, V. B. (1989). *Spiritual dimensions of nursing practice.* Philadelphia: W. B. Saunders.

Cohen, A. A., & Mendes-Flohr, P. (1987). *Contemporary Jewish religious thought.* New York: Free Press.

Edgar, M. A. (1989). *An investigation of the relationship between nurses' spiritual well-being and attitudes of their role in providing spiritual care.* Unpublished master's thesis, Department of Nursing, California State University, Los Angeles.

Ellis, J. R., & Nowlis, E. A. (1985). Values and beliefs. *Nursing: A human needs approach.* Boston: Houghton Mifflin.

Ellison, C. W. (1983). Spiritual well-being. Conceptualization and measurement. *Journal of Psychology and Theology, 11,* 330-340.

Ellison, C. W., & Smith, J. (1991). Toward an integrative measure of health and well-being. *Journal of Psychology and Theology, 19*(1), 35-48.

Emblen, J., Fitchett, G., Farran, C., & Burck, J. (1991). Identifying parameters of spiritual need. *The CareGiver Journal, 8*(2), 44-49.

Emblen, J. D. (1992). Religion and spirituality defined according to current use in nursing literature. *Journal of Professional Nursing, 8*(1), 41-47.

Emblen, J. D., & Halstead, L. (1993). Spiritual needs and interventions: Comparing the views of patients, nurses, and chaplains. *Clinical Nurse Specialist, 7*(4), 175-182.

Farran, C., Fitchett, G., Quiring-Emblen, J., & Burck, J. (1989). Development of a model for spiritual assessment and intervention. *Journal of Religion and Health, 28*(3), 185-194.

Fowler, J. (1981). *Stages of faith: The psychology of human development and the quest for meaning.* San Francisco: Harper & Row.

Grove, P. (Ed.). (1986). *Webster's third new international dictionary, unabridged.* Springfield, MA: Merriam-Webster.

Kim, M. J., McFarland, G. K., & McLane, A. M. (Eds.). (1987). *Pocket guide to nursing diagnoses* (2nd ed.). St. Louis, MO: C. V. Mosby.

Kohlberg, L. (1968). Moral development. *International Encyclopedia of Social Science*. New York: Macmillan-Free Press.

Lambert, C. E., & Lambert, V. A. (1987). Hardiness: Its development and relevance to nursing. *Image: Journal of Nursing Scholarship, 19*(2), 92-95.

Mansen, T. J. (1993). The spiritual dimension of individuals: Conceptual development. *Nursing Diagnosis, 4*(4), 140-147.

Maslow, A. (1970). *Religions, values, and peak-experiences.* New York: Viking.

Moberg, D. O. (1984). Subjective measures of spiritual well-being. *Review of Religious Research, 25,* 351-364.

Murray, R., & Zentner, J. (1985). Religious influences on the person. *Nursing concepts for health promotion* (3rd ed.). Englewood Cliffs, NJ: Prentice Hall.

New York Bible Society International. (1978). *The Holy Bible, New International Version.* Grand Rapids, MI: Zondervan.

Olthuis, J. H. (1989). On worldviews. In P. Marshall, S. Griffioen, & R. Lieuw (Eds.), *Stained glass: Worldviews and social science.* New York: University Press.

Pumphrey, J. B. (1977). Recognizing your patients' spiritual needs. *American Journal of Nursing, 77,* 64-70.

Reed, P. (1992). An emerging paradigm for the investigation of spirituality in nursing. *Research in Nursing and Health, 15,* 349-357.

Runia, K. (1963). *I believe in God.* Chicago: Intervarsity Press.

Springer, S. P., & Deutsch, G. (1981). *Left brain, right brain.* New York: Freeman.

Stallwood, J., & Stoll, R. (1975). Spiritual dimensions of nursing practice, Part C. In I. Beland & J. Passos (Eds.), *Clinical nursing* (3rd ed., pp. 1086-1098). New York: Macmillan.

The Westminster Confession of Faith. (1643-1648). Philadelphia, PA: Great Commission Publications.

8

The Family as Client

An Interactive Interpersonal System

Darlene E. McCown
Barbara M. Artinian

Family Biological Subsystem
Family Psychosocial Subsystem
Family Spiritual Subsystem
Family Clinical Example

Although most nursing care is directed toward the individual, the Intersystem Model provides a format for assessing and intervening with the family as a whole. Because it is possible to think about a member of a group at one level and think about the group itself at a more abstract or second level, it is possible to consider individuals to be members of a family and also the family itself (Watzlawick, Weakland, & Fisch, 1974). This makes it possible to provide nursing care to the individual and, at another hierarchical level, to the total family system, recognizing that change in an individual is a different type of change from change in a family group.

The family can be defined as a system of interacting persons who live together over time developing patterns of kinship and who hold specific role relationships to each other, characterized by commitment and attachment, and who have economic, emotional, and physical obligations to each other. This definition includes the traditional two-parent family as well as alternative family forms. Stuart (1991) describes the characteristics of the family system as follows:

> As interactional systems, families operate according to rules and principles that apply to all systems, including interdependence, interrelatedness, nonsummativity, circular causality, equifinality, and homeostasis. (p. 31)

Interaction occurs within the system and between the system as a whole and the environmental context. The family system occupies space and takes on unique forms and personality necessary to regulate its health care activities and develop a sense of coherence. Through interaction, the family develops its own unique paradigm, which Reiss (1981) defines as "the enduring, fundamental, and shared assumptions families create about the nature and meaning of life, what is important, and how to cope with the world in which they live" (cited in Burr, 1991, p. 439). This shared paradigm allows pattern and predictability to occur in family processes, which enables families to decide how to allocate resources to attain family goals.

Just as the individual has subsystems, so does the family. The family system may be viewed as composed of three main subsystems: biological, psychosocial, and spiritual. When the family encounters a stressor, the stressor is processed through these subsystems. In the biological subsystem, the stressor interacts with family characteristics such as developmental stages, gender composition, role function, genetic factors, illness trajectory, age-related health changes, and availability of food, shelter, and medical care. In the psychosocial subsystem, the stressor is influenced by family organizational patterns as evidenced by family themes, marriage forms, family type, coping strategies, family strengths, family liaisons or coalitions, culture and ethnicity, decision-making patterns, parenting skills, family history of rituals and traditions, and family secrets. These reflect the cognitive, affective, and interactional climate of the family. The stressor is also processed through the spiritual subsystem in which the family belief system, the religious values, the religious practices, and the affective responses are demonstrated by church affiliation, spirituality, the development of faith, and worship. These characteristics are summarized in Table 8.1.

Output at the family level can be measured as the level of health knowledge about the particular problem, the orientation to health care, and the health practices and skills exhibited by the family. The characteristics outlined previously mediate health stressors experienced by the family and can be seen as generalized resistance resources or coping resources in the salutogenic model developed by Antonovsky (1987).

In the salutogenic model, it is proposed that individuals, over time, develop a global orientation to life that expresses the extent to which they see the world

Table 8.1 Family Subsystems

Biological Subsystem	*Psychosocial Subsystem*	*Spiritual Subsystem*
Developmental stage	Family themes	Church affiliation (beliefs)
Gender composition	Marriage forms	Spirituality (values)
Role function	Family type	Faith development (practices)
Genetic factors		
Illness trajectory	Coping strategies	Worship (affective response)
Age-related health changes	Family strengths	
Availability of food, shelter, and medical care	Family liaisons (coalitions)	
	Culture and ethnicity	
	Decision-making patterns	
	Parenting skills	
	Family rituals and traditions	
	Family secrets	

as comprehensible, manageable, and meaningful. As an ongoing system, the family also develops a global orientation to life. In an experience of stress, the family's appraisal of the overall situation leads to the actual response of the family to the stressor and can be measured as the family situational sense of coherence (SSOC) scored as high, medium, or low. In the general systems approach, this measurement of the family reality can be considered as the output and expresses the actual coping ability of the family system.

This chapter addresses selected theories and concepts underlying the three family subsystems that compose the family system in the Intersystem Model. An understanding of these theories will enable the nurse to more adequately assess family SSOC to assist families to plan for their health care needs. An example from a community health case study will illustrate the process of providing care to a family system using the Intersystem Model framework.

Family Biological Subsystem

Family Gender Issues

Traditionally, families have included both males and females. As families engage in the functions necessary to keep the system viable and healthy, roles and activities are often distributed according to gender. Usually by age 3 or 4 years, children become socialized into roles according to gender and the culture of gender differences develops unnoticed.

Stereotypical traits have been identified for each sex (Allensworth & Byrne, 1982; Bem, 1974; Chodorow, 1978; Gilligan, 1982; Grusec & Lytton, 1988; Maccoby & Jacklin, 1974). Masculine traits typically include being aggressive, assertive, competitive, active, independent, stoic, able to exercise justice, visual-spatial in perception, mathematical, ambitious, and dominant. Feminine traits often include being dependent, sensitive, passive, emotional, verbal, relationship oriented, fragile, and drawn toward children. It has been documented that males are more likely to engage in risk-taking behaviors and women to practice health-promoting and prevention activities (Nathason, 1977).

Another commonly identified gender issue of importance to nurses is related to the control of emotions. Males are expected to restrain from expressing emotions of fear and sadness. The constraints placed on emotional expression in males may result in lack of buildup of energy and contribute to stress-related illness and activities found in men such as increased alcohol consumption, violence, and cardiac disease. Likewise, for women the stress of social devaluation, lack of authority, discrimination, and the possibility of "learned helplessness" (Seligman, 1981) are thought to contribute to inward-type disorders such as depression, anorexia, and dependence on tranquilizers. It is possible that forcing individuals into sex-stereotyped roles and expectant behaviors may violate individual needs and bring stress and discomfort to the family member.

The impact of sex differences has received particular attention from Gilligan (1982). The standard of interpreting personality attributes on the basis of a male-dominated model has been challenged as gender bias. Sound reasoning and careful research reveal what Gilligan labels a "different voice" for women in regard to moral reasoning in particular. Women view moral decision making from an interpersonal connectedness orientation based on sensitivity and responsibility to people. They feel a need to alleviate trouble and problems in the world. Males, in contrast, tend to organize social relationships in hierarchical order and value respect and protection for the rights of others and obedience to rules and principles. Females tend toward an interpersonal or caring orientation. Both perspectives have value, and neither is superior to the other. There is some research, however, that supports the idea that rigid sex typing is associated with less favorable personal adjustment, especially for women (Brewer & Blum, 1979; Hoffman & Fidell, 1979; Jones, Chernovetz, & Hansson, 1978). It has also been hypothesized by Feldman (1982) that sex role typing contributes to dysfunctional actions and attitudes that interfere with positive parenting for both men and women. Fathers exhibit a tendency toward authoritarianism and rigidity, and mothers appear overprotective and overcontrolling.

Nursing Assessment

Awareness of common gender stereotypes is helpful to the nurse when assessing the family. Knowing the sex of family members offers clues to potential health problems such as depression and to concerns regarding relationships versus justice and actions. Accurate assessment of sex role types within a particular family can assist the nurse in gaining insight regarding motivation and behavior of various individuals and in forming a basis for effective communication and a positive working relationship.

Family Developmental Theory

For more than 40 years, family developmental theory has taken a life cycle approach toward understanding how families function and relate over time. In this conceptual frame of reference or general orientation, families pass through obvious stages over the "life course" (Aldous, 1990). The functions commonly attributed to the family development perspective have changed over time and currently focus on the emotional support and socialization needed for living in the extended environment. These functions include

1. Giving affection
2. Providing personal security, nurturance, and respite
3. Giving satisfaction and purpose to life
4. Assuring continuity of companionship, acceptance, and communication
5. Providing socialization and status
6. Teaching values and beliefs (Duvall & Miller, 1985; Janosik & Green, 1992)

The life course perspective best represented by the family development theory of Duvall and Miller (1985) follows the family across hierarchical and universal stages. The stages are based on the establishment of a marital or couple relationship and then follow the birth and transition events of the first child. Although Duvall and Miller record 8 stages, others using a family developmental approach identified as many as 24 stages (Rodgers, 1962). Duvall and Miller have identified four factors pertinent to the family life cycle stages: (a) plurality patterns, (b) age of oldest child, (c) school placement of oldest child, and (d) functions and status of families before children come and after they leave the home.

The family developmental perspective recognizes the importance of structural roles and positions within the family. At each advancing stage in the cycle, an ever-changing number of roles are added. The family begins as a couple with the roles of husband and wife, among others, and expands at its peak to include roles of wife, mother, husband, father, son, daughter, sister, brother, grandmother, grandfather, aunt, and uncle. This conceptualization of the family development readily shows the variation of the family over time and reveals the significant amount of time the couple spends in a dyad without children in the home.

Table 8.2   Divorced Family Developmental Cycle

Stage/Steps	Emotional/Developmental Tasks
Decision to divorce	Acceptance of inability to resolve marital conflicts
	Acceptance of one's own part in the failure of the marriage
Planning the breakup of the marriage	Supporting workable arrangements for all parts of the family
	Working cooperatively on problems of custody, visitation, and finances
	Dealing with extended family and friends
Separation	Mourn loss of intact family
	Maintain cooperative coparental relationship
	Resolve attachment to spouse and restructure relations with extended family
	Adapt to living separately and restructure parent-child relationships
Divorce	Overcome feelings of hurt, anger, guilt, and so on
	Relinquish hope for reunion and expectation of marriage
	Maintain contact with extended families
Entering new relationship	Recover from loss of first marriage and establish openness in new relationship
Decision to remarry	Accept one's own fears and those of new partner and children about forming a new family
	Accept need for time for adjustment to (a) new roles; (b) boundaries of time, memberships, space, and authority; and (c) affective issues of guilt, loyalty, conflicts, mutuality, and past hurts
Remarriage and establishing a new family	Resolve attachment to previous spouse and ideal intact family
	Accept a new model of family with permeable boundaries
	Establish attachment to new partner and inclusion in family system
	Realign relationship throughout subsystems to permit interweaving of several systems
	Share memories and histories across systems

SOURCE: Combined and adapted from McGoldrick and Carter (1982) and Ransom, Schlesinger, and Derdeyn (1979).

Although Duvall and Miller (1985) focused on the intact family, others have studied the family life cycle events of separating or divorced families (McGoldrick & Carter, 1982). A somewhat different pattern of expansion, contraction, and realignment of relationships, additional stages, and tasks faces separated families, as shown in Table 8.2. Neither of these examples adequately considers the many other variations of families relative to culture, poverty, and nontraditional forms of marriage. Studies on female-headed and black families have been reported (Hill, 1986; Mattessich & Hill, 1987).

The family developmental life course framework builds on several other main features. It assumes the members interact within the unit and within the external environment on the basis of roles. The members fulfill specified tasks pertinent to the life of the family whether it is in the expanding or contracting cycles. Developmental tasks are those that arise at a specified time in an individual's life when, if accomplished successfully, they will contribute to success and advancement to later tasks. Although the developmental task originates from within the system, social and cultural pressures encourage appropriate developmental tasks. At times within the family, developmental tasks of various subsystems may be congruent or in conflict. The teenager's task is to establish independence. The parent, however, may still feel a need for close supervision and ongoing parental control, thus resulting in continuing struggle between them. As the family makes transitions and moves from one stage to the next, these moves are attended by a new set of tasks and role expectations. Successful accomplishment of the tasks and goals brings a sense of confidence, satisfaction, and success to the family (Rodgers & White, 1993).

The family development cycle has changed over time in response to multiple factors. Increased mobility has distanced generations of families and altered the space dimension. Delayed childbearing has extended the first stage. Decreasing family size has reduced the number of relationships required in the "normal" family. Increasing rates of divorce, however, have complicated parenting roles and required alternative household arrangements. Longevity frequently enables extended families to go to the fourth generation and has added responsibilities for care of elderly members.

Nursing Assessment

The developmental framework aids in understanding and assessing the family as it continues over time. A family developmental perspective views the changes in growth, development, dissolution, or decline across the entire family life cycle. Family development conceptualization allows nurses to predict and anticipate the next phase and probable events in the family life cycle. This ability is dependent on knowledge in three areas: (a) the place of the family in time and its life cycle; (b) the number, ages, and relationships of household members; and (c) the community identification of the families' ethnic, religious, and social status (Duvall & Miller, 1985). The family developmental orientation serves nurses well in understanding the family system as a whole as it experiences significant events that customarily intersect with the health care system such as birth, significant illness, and death.

Family Psychosocial Subsystem

Family Structures and Themes

The family as a system of interacting individuals is an open, semipermeable system. The family exchanges energy internally within its subsystems and also with the external environment and changes and adapts according to the forces involved. Kantor and Lehr (1975) described three basic structural arrangements or family types for regulating distance and processing information and adapting to change. The major family types are open, closed, and random.

Open, closed, and random family types can be distinguished by the kind of boundaries they form. Boundaries are defined as the rules and norms within the family that develop over time and guide family behaviors. Individual boundaries serve to establish and protect the self of individual members. Interpersonal subsystem boundaries may be established, for instance, between spouses to protect the parenting system. Boundaries—the rules defining who participates and how—surrounding the family and its subsystems need to be strong enough to protect the systems but also permeable enough to allow an exchange of information between the individual members and the family within the community. If boundaries between members are weak, the development and sustenance of individuality is threatened and members become enmeshed. When boundaries between subsystems become rigid, such as may occur between parents and children, disengagement from each other may result (Janosik & Green, 1992). Family boundaries also determine the amount and type of contact with the world outside the family system. It is important for the professional who works with families to distinguish when a person is responding from a personal (self) system perspective or from a group member or family perspective. The family protects its boundaries by developing strategies for receiving input and giving out to the environment. These strategies are the key to developing and maintaining the family's functioning and focus on access and target dimensions.

Each family type presents unique features that maintain it as a system. Strategies are used by the family to help to interpret, protect, and maintain family traditions. Closed-type families establish fixed, constant feedback systems. The core purpose of the closed family is to create stability in the family process. These families admit only information, events, and ideas that are congruent with their own norms, values, and beliefs. They keep external traffic and input to a minimum. This type of family may be identified by external evidence of tight security systems and rigid scrutiny of visitors and credentials. Pace in closed families is fixed. Reading and viewing materials and school activities are carefully evaluated and controlled. In closed families,

Table 8.3   Kantor and Lehr's Family Types and Access and Target Dimensions

Dimension	Family Type		
	Closed	Open	Random
Goal	Stability through tradition	Adaptation through consensus	Exploration through intuition
Access distance regulation			
Space	Fixed	Moveable	Dispersed
Time	Regular	Variable	Irregular
Energy	Steady	Flexible	Fluctuating
Target distance regulation			
Affect (joining-separating)	Durability	Responsiveness	Rapture
	Fidelity	Authenticity	Whimsicality
	Sincerity	Latitude	Spontaneity
Power (freedom-restriction)	Authority	Resolution	Interchangeability
	Discipline	Allowance	Free choice
	Preparation	Cooperation	Challenge
Meaning (sharing-not sharing)	Certainty	Relevance	Ambiguity
	Unity	Affinity	Diversity
	Clarity	Tolerance	Originality

SOURCE: From Kantor and Lehr (1975). *Inside the Family* (p. 150). Used with permission from Jossey-Bass.

interactions of family members are governed by authority figures. Kantor and Lehr (1975) identify various other ways in which families control access to space, time, and energy (Table 8.3).

The open-type family purpose is to create a system that is adaptive to the needs of the individual and society. Emotions and affection are overtly expressed. Consent of members is sought in decision making. Persuasion rather than coercion is a mode of operation. Autonomy and negotiation are encouraged and cooperation is fostered. New ideas are incorporated into the system. Communication in open families is authentic and informal.

The random family as described by Kantor and Lehr (1975) is identity seeking. Reasoning is predominantly inspirational and intuitive. Variant views and ambiguity are acceptable. Communication is flexible and variant. Creativity is encouraged and valued. Diversity and originality are ideals of the random family system.

Just as family systems have typical patterns of response, so do the individual members. Kantor and Lehr (1975) describe four basic parts that individuals in the family play: mover, follower, opposer, and bystander. The initiator of the action is the mover. Those who respond are comovers. Followers agree with the initiator. Opposers challenge the action of the mover. Bystanders watch the action but neither agree nor disagree. Individuals may take on more than one part at a time, and more than one person can play a part at the same time. Identification of the parts various members play in a family system as a rule and in various circumstances offers insight and direction for nurses working with families.

Each member and family type has flaws and strengths. In reality, families do not rigidly follow a single type and may mix strategies across types. Families tend toward certain predominant strategies and type, however, as they seek to adapt and maintain family stability through balancing the forces of equilibrium (or relaxation), disequilibrium (tension), and the restoration of equilibrium (relaxation of tension). Tension in the family helps to generate energy to foster growth. Excessive tension, however, may strain the system and requires mechanisms for discharge of tension from the family. Health care professionals can help families discover and find strategies for sustaining balance and discharging ongoing tensions.

Dysfunctional Family Characteristics

On the basis of the typical family profiles and system responses of families having children with emotional problems, Minuchin, Rosman, and Baker (1978) suggested five characteristics to help the professional recognize families with psychosomatic problems. The characteristics include

1. Enmeshment—members are overinvolved with one family member's situation or in the relationship between two members.
2. Overprotectiveness—members show a higher degree of concern for each other's welfare than is warranted by circumstances.
3. Rigidity—members engage in patterns of interaction that are repetitive and inflexible to maintain the status quo.
4. Conflict avoidance—members do not openly display disagreement despite obvious conflict.
5. Child as regulator—the child's symptom becomes the mediator of parental conflict (pp. 30-32).

Nursing Assessment

The nurse can be aided in the identification of member and family types by attending to specific behaviors. Is the nurse readily admitted into the home? How freely is personal information shared? Who is present at the interview? Is the home securely locked and closed up? Where in the home does the interview take place—in the formal living room, kitchen area, or bedroom?

Who makes suggestions, speaks, and makes requests in the family? Who carries out requests? What are the family rules, and how are they obeyed? The primary goal for the nurse is to assist the family in developing systems with the integrity and the flexibility to adapt to the changing needs of the system to maintain itself and meet its unique goals.

Family Stress Theory of Adjustment and Adaptation

Nearly 50 years ago, the basics of family stress theory were first introduced by Hill (1949). Hill attempted to explain the response of families to stressor events represented by war separation and reunion. He developed a basic pattern described as the ABCX model of family stress. Traditionally, A is defined as the stressor event, B is the family crisis-meeting resources, C is the definition the family makes of the event, and X represents the crisis.

Recent scholars have refined and expanded the family stress model (Burr, 1973; Hansen & Johnson, 1979; McCubbin & McCubbin, 1987; Patterson & Garwick, 1994). The importance of family types, family strengths and capabilities, and recognition of the "pile up" of stressor demands on the family have been added to the model. It has been reformulated as the T-Double ABCX Model by McCubbin and McCubbin (1987). This two-phase model of family adjustment and adaptation is based on the following assumptions:

1. Families face hardships and change as a part of life.
2. Families develop coping strategies to deal with disruptions and to promote growth and development in the family.
3. Strengths are developing to protect families from unexpected events.
4. Families benefit from and give to the community network of relationships and resources.

Phases

Adjustment Phase. The first adjustment phase of the model suggests that there is a stressor event (A) that interacts with the family vulnerabilities (V), which are partially determined by a pile up of demands (stressors, transitions, strains, and life cycle stage). These factors then interact with the family typology (T), including the types such as regenerative, resilient, rhythmic, and balanced. The interactions continue depending on the family's resistance resources (B) interacting with the appraisal the family makes of the event (C) and interacting with the family's problem-solving and coping responses (PSC) to the situation. This series of events results in the family's adjustment response (X) to the stressor event. Families may make a favorable or unfavorable adjustment.

If it is a maladjustment response, the family is determined to be in crisis and, according to McCubbin and McCubbin (1987), a process of adaptation to the crisis then occurs. The first phase of the model can be diagrammed as follows:

$$A \rightarrow V \rightarrow T \overset{C}{\underset{B}{\leftrightarrow}} PCS \rightarrow X$$

Adaptation Phase. A similar pattern is described in the family adaptation phase to the crisis situation (McCubbin & McCubbin, 1987). The T-Double ABCX model of adaptation describes the family's adaptation (XX) or transition back into crisis in response to a stress situation by the pile up of demands (AA) interacting with the family's level of regenerativity (R) interacting with the family typology (T). This interacts with the family's strengths (BB; adaptive strengths, capabilities, and resources) interacting with the family's appraisal or meaning attached to the situation (CC) and the family's schema or view of the events (CCC). Social support from friends and community (BBB) contributes to the family's PSC and the level of adaptation to crisis. The family adaptation phase of the T-Double ABCX model is diagrammed in Figure 8.1.

Pile Up. The pile up of stressors or demands placed on the family is generated over time. The demands come from multiple sources including normal developmental transitions such as birth, marriage, and death. Another aspect of demand comes from the specific hardships imposed by the stressor itself such as loss of sleep following birth of a child or during hospice care of a dying family member. Prior strains represent a third aspect of the pile up of demands. In the event of a new stressor, dealing with prior strains as well may contribute to an overall source of stress. Efforts by the family to cope with and manage the crisis situation may also contribute to the pile-up demand. For example, remodeling a house to make room for a new member brings its own set of problems. A final source of stress contributing to pile up is the intrafamily and social ambiguity that is part of crisis situations. Family and social roles, boundaries, rules, and responsibilities may be unclear in times of change.

Family Strengths and Resources

Family strengths and capabilities are an important feature of Family Stress and Adaptation Theory. In this model, capability is defined as the potentiality the family has available for meeting demands (McCubbin & McCubbin,

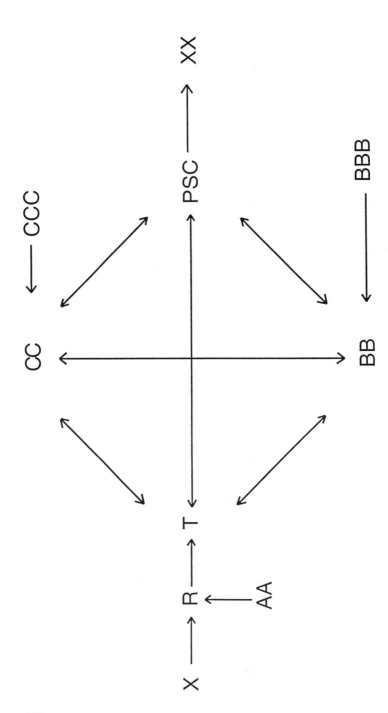

Figure 8.1. Outline of Adaptation Phase of T-Double ABC X Model.

SOURCE: From McCubbin and Thompson (1987). *Family Assessment Inventories for Research and Practice* (p. 15). Used with permission from the Family Stress Coping and Health Project.

1987). Capabilities are of two types: resources and strengths (what the family has) and coping behaviors and strategies (what the family does using the coping behaviors).

Personal resources available to families in managing stress include factors such as innate intelligence; knowledge; skills and training; personality traits such as organizational and social skills; physical and emotional health; sense of mastery and control; and self-esteem. The erosion of self-esteem through the effect of chronic strain on the family's inability to achieve successful mastery finds linkages with poor adaptation, such as depression, to family stress.

Family system resources are also described by McCubbin and McCubbin (1987) as family strengths. Cohesion, which is defined as emotional bonds of unity shared with family members, is one major strength force. Cohesion includes family attributes such as trust, appreciation, support, integration, and respect for individuals. Adaptability or the family's capacity to change and shift course is another well-documented family strength. Another family resource is family organization. This resource is characterized by shared parental power, clear boundaries and roles, and rules (Kantor & Lehr, 1975). Communication patterns that are clear, warm, and consistent enable families during stressful situations. These patterns were identified by Curran (1983) as the most frequent trait of a healthy family.

Families in crisis situations need resources beyond their personal capacities. Community resources available to assist families vary from place to place and include services such as health care, shelter, food, spiritual care, and financial and legal assistance, as well as a long list of other specific resources. Social support has been identified as a significant resource for families during crisis events. McCubbin and McCubbin (1987) have provided the following helpful definition and description of social support as

> information exchanged at the interpersonal level which provides a) emotional support, leading the individual to believe that he or she is cared for and loved; b) esteem support, leading the individual to believe he or she is esteemed and valued; c) network support, leading the individual to believe he or she belongs to a network of communication involving mutual obligation and mutual understanding; d) appraisal support, which is information in the form of feedback allowing the individual to assess how well he or she is doing life's tasks; and e) altruistic support, which is information received in the form of good will from others for having given something of oneself. (p. 19)

Nursing Assessment

Nurses working in acute, chronic, and community settings frequently have direct contact with families in the struggles of crisis adjustment and adaptation. Through the assessment process, the nurse can help the family identify the pile up of demands placed on the system. The nurse can recognize those most vulnerable to or able to regenerate from the various stressors. The nurse may intervene to help the family relieve some of the demands and strengthen personal or community resources utilization. The nurse may identify ways for the family to strengthen and clarify boundary and role issues or to improve lines of communication and power between members or between the family and community systems. The advocacy collaborative role of the nurse is especially key in facilitating intersystem linkages for families. Skills in communication and social support as well as knowledge of formal family assessment tools, such as the Family Adaptability and Cohesion Scales, Family Inventory of Resources Management, Feetham Family Functioning Survey Manual (Feetham, 1991), and Family Environment Scale, are essential for nurses who provide health care to families.

Family Spiritual Subsystem

Development of Faith

It is generally acknowledged that the family plays a major role in the religious socialization process. Although religious variables, such as denominational preference and frequency of church attendance, frequently have been included in family research, attention to the transmission of and the main effects of spiritual beliefs is just emerging (Boss, Doherty, LaRossa, Schumm, & Steinmetz, 1993; Hayes & Pittelkow, 1993). Developmental theorists, however, have studied faith as it develops across the life cycle (Fowler, 1981, 1984; Stokes, 1983). An understanding of the stages of faith provides a basis for nurses to appropriately support and strengthen family members' coping strategies during crisis situations involving health care issues.

The stages of faith development have been carefully described by James Fowler (1981, 1984) from a structural perspective. On the basis of hundreds of personal interviews with people 4 to 80 years old, he identified six discrete and hierarchical stages of faith. The age ranges for the stages are arbitrary points and reflect general patterns of faith development. The stages depend on age and maturation in the sense that these factors provide some of the conditions necessary for stage transition (Fowler, 1980). They are not sufficient conditions. Faith can be arrested at any stage. The idea that factors such as environmental stimulation and suitable models for imitation significantly influence the timing and movement through faith stages is supported by the research of Hayes and Pittelkow (1993). The six stages of faith are

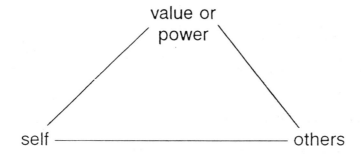

Figure 8.2. Triad of Faith

0. Undifferentiated faith—infancy
1. Intuitive-projective faith—early childhood
2. Mythic-literal faith—school years
3. Synthetic-conventional faith—adolescence
4. Individuative-reflective faith—young adulthood
5. Conjunctive faith—mid-life and beyond
6. Universalizing faith

The defining characteristics of each stage have been described in Chapter 3.

"Faith" for Fowler does not refer to a specific set of beliefs or religious dogmas. It is the human activity by which one understands the self in relation to the ultimate concerns of life (Fowler, 1981). The contents of faith include (a) centers of value, (b) images of power, and (c) master stories. Faith is a triad incorporating self, others, and values or power. This concept is illustrated in Figure 8.2.

Conversion entails a change in the contents of faith. A person experiencing conversion establishes a new or renewed value system. There is a new or fresh adherence to the narrative accounts or stories of faith. New relationships and attitudes toward others, self, and the ultimate authority are formulated. These changes reflect a conversion experience.

Nursing and the Spiritual Subsystem

Nurses have contact with families at points in time when critical transitions are taking place such as birth and death. By accurately identifying the family members' belief system and the ultimate source of power and strength for them, the nurse can assist them to maximize their inner and external resources, thus increasing their SSOC. Encouraging religious rituals, such as prayer, communion, and anointing with oil, serves to enhance family coping resources. Referrals to religious groups and clergy can bring external resources into the health care system to help support and stabilize families during times of need. The nurse's role as an educator can help parents recognize their key role in transmitting faith to their children and reinforce the importance of meaning, religion, and values in strengthening family systems.

Family Clinical Example

The following is an excerpt selected from a 6-week case study written by Beverly Brownell, RN (a graduate student at Azusa Pacific University),[1] that illustrates how a nurse and family system can develop a mutual plan of care using the Intersystem Model framework.

Focus of Situational Interaction

Mr. F is at home receiving hospice care through a home health agency. The nurse visited to check on pain control and care of the decubitus ulcer. The ulcer was not healing, and Mrs. F was concerned that it was her fault.

Assessment

Developmental Environment

Biological Subsystem

Mr. F has had prostate cancer for the past 2 years, for which he has had radiation therapy. For the past 6 months, he has been treated with Lupron injections without success. He also has associated bone metastases, particularly in the spine and knees. He is a diabetic controlled by diet and has a fungating stage II decubitus ulcer on his coccyx that requires dressing three times a day. He also has a Foley catheter because he has a tendency to go into bladder retention. He has a subcutaneous morphine infusion administered by pump to control his pain.

Psychosocial Subsystem

The family unit consists of Mr. F, who is 70 years of age, Mrs. F, who is Korean and in her late twenties, and Mrs. F's niece, who lives with them but does not play an active role and prefers to keep her distance. Mr. and Mrs. F have been married for 5 years and live in a large four-bedroom house in a fairly affluent neighborhood. Mr. F has two sons by his first marriage, one of whom lives a block away and visits regularly. That son appears to get on relatively well with Mrs. F. The other son lives in Denmark and visits infrequently. He was opposed to the marriage. Mr. F also has a brother in the area who visits on occasion. Mr. and Mrs. F have what appears to be a functioning marriage in which Mrs. F plays a rather subservient role.

Spiritual Subsystem

The family does not appear to have any strong religious affiliation although Mrs. F did say that she used to go to church regularly in Korea.

Situational Environment

Intrasystem Analysis of the Family

Detector Function. Mrs. F did not know how to do the dressing changes or how to turn her husband to relieve pressure on his coccyx.

Selector Function. Mrs. F is deeply committed to her marriage and to caring for her husband but is overwhelmed by the responsibility. She wants to provide physical care for her husband. They both see him as too young to be dying, but Mrs. F is more realistic.

Effector Function. Mrs. F is very competent in caring for her husband but is hesitant because she does not know what to do. She is supportive and eager to please him. She is keen to learn and learns very quickly. Mr. F tries to be cooperative and his attitudes are sound, but he evidences a demanding and manipulative personality that is often disguised by humor.

Intrasystem Analysis of the Nurse

Detector Function

The nurse has worked in a medical-surgical ward and has had oncology experience. She is knowledgeable about Mr. F's multiple problems. She knows the various methods of treating decubitus ulcers. She understands the learning abilities of the elderly and the effect of loss of control and the fear of death on the patient and family.

Selector Function

The nurse values and respects the organizational pattern of the family and their belief system. She realizes that it is necessary to gain the confidence of the family by demonstrating her competence and caring in assisting with the problem of the ulcer before they trust her to assist them with the larger issues of dying.

Effector Function

The nurse has a positive attitude and is able to differentiate between realistic and nonrealistic solutions to problems. She is able to organize interventions as they are appropriate.

Assessment

Identification of Stressors

Metastatic cancer with a life expectation of less than 6 months
Poor family support system to assist Mrs. F
Nonhealing decubitus ulcer
Pain

Coping Resources

Supportive marital relationship
Strong motivation to learn and participate in rehabilitative process
Strong financial resources

Scoring on Situational Sense of Coherence

Comprehensibility = 1. The family understands much about the effect of cancer but does not know how to manage the daily care that is required. Mrs. F does not know how to dress the wound site or position her husband.

Meaningfulness = 2. Mrs. F is willing to learn how to provide care for her husband but is overwhelmed because she has not yet had time to adjust to the terminal diagnosis. Everything had happened too quickly for them to understand or assimilate. Mr. F is still placing hope for recovery on treatments that are not effective rather than focusing on the daily care he needs.

Manageability = 2. The family has agreed to accept the services of the hospice nurse, a home health aide, and a social worker. There are adequate financial resources to provide for the supplies and the services he needs.

Nursing Diagnoses

(a) Impaired skin integrity related to effects of immobility as evidenced by stage II decubitus ulcer on sacrum, (b) knowledge deficit about daily care related to recent onset of severe physical problems, (c) alteration in family process related to anxiety about impaired physical condition indicating terminal cancer, and (d) high risk for caregiver role strain related to the burden of caregiving.

Nursing Goals

Because both Mr. and Mrs. F were overwhelmed with the situation, we agreed that each issue needed to be prioritized and dealt with one at a time. We were now more comfortable with each others and there were signs of the beginnings of a relationship among the three of us. Therefore, I thought it was the appropriate time to begin to teach Mrs. F how to do dressing changes. We agreed that small goals would be the best way to deal with the situation. Therefore, the following immediate goals were set:

1. Mrs. F will learn to do the dressing change. Rationale: Mrs. F will be less anxious if she knows how to provide care.
2. Correct positioning to relieve pressure on the sacrum will be carried out. Rationale: Ulcer wound is not healing as it should and could require hospitalization if an adequate program is not instituted.

Intersystem Interaction

Communication of Information

The importance of a turning regimen that would take pressure off the sacrum was emphasized. What to look for in assessing the wound site was discussed, and a demonstration of wound care was given.

Negotiation of Values

The importance of the turning regimen was accepted. Because the family was not ready to discuss the larger issues of prognosis, the focus was on caregiving issues to build up the relationship and allow time for them to absorb the fact of his terminal condition.

Organization of Behaviors: Mutual Plan of Care

Mrs. F gave a return demonstration of wound care that was accurately done. We tried many ways to position Mr. F with numerous pillows in different arrangement, but despite all our efforts Mr. F was not at all comfortable in any of the new positions because of his bone metastases. We finally agreed to order an alternating air mattress and have his whole body supported. Written instructions were left in the home for reference along with the necessary supplies. Mrs. F appeared to have questions that she was uncomfortable broaching in the presence of her husband. When she followed me to my car, I asked her how she was really coping with everything. Her body language indicated that she was in a great deal of turmoil. The following plan was agreed on to accomplish before the next visit:

1. Mrs. F would order an air mattress.
2. Mrs. F would do dressing changes and follow the agreed-on turning procedure, turning him from side to side as long he could comfortably tolerate it.
3. Mrs. F would prepare all the questions she needed answers to that she did not feel ready to discuss with her husband so that we could discuss them on the next visit.

Ongoing Intersystem Interaction

Because this is a long-term interaction between the nurse and family, evaluation of the effectiveness of the interventions was done at weekly intervals. The following intersystem interaction regarding these issues occurred at the next visit. Using the Intersystem Model process of assessing family and nurse knowledge, values, and behaviors (the detector, selector, and effector functions) and communicating information, negotiating values, and organizing behaviors (intersystem interaction), the nurse worked with the family during the visit. Mr. F said that the ulcer on his coccyx was causing him a great deal of local discomfort.

On inspection of the wound site, there seemed to be no improvement in condition. The ulcer now had a necrotic center about 2 cm in diameter. Both Mr. and Mrs. F still expressed a willingness to do whatever was necessary to promote healing of the site. Mr. F, however, was not at all compliant in maintaining the turning schedule to reduce pressure to the site. The alternating air mattress was in place and was improving the comfort. We all agreed, however, that some means had to be arranged to get as much pressure off the area as possible. We tried different ways to place the pillows and finally agreed on a possible solution, which was immediately put into place. With Mr. F now settled, Mrs. F and I migrated downstairs to the kitchen. Having observed how emotional she had been upstairs and after getting her to describe a typical day, it was not difficult to see that she was already feeling the strain of caregiving. In addition, she obviously had many unresolved questions to ask. I attempted to discuss Mr. F's terminal condition in a fairly direct manner without appearing too blunt, because I sensed this was very difficult for Mrs. F. She expressed that she was finding it difficult to assimilate the fact that her husband, who had always been such a healthy and vibrant man, was now dying. She expressed that she was just not ready to let him go yet and needed to know if there was anything she could possibly do. We discussed Mr. F's prognosis at length, but I could see that Mrs. F was not taking in what I had to say and that it was going to take a lot of support from her family and me to get her through her anticipatory grief to the point of acceptance. We finally ended the visit by agreeing to continue with our conversation in stages. Support systems were explored, such as more family involvement from Mr. F's brother and son in the evening, volunteer and homemaker services, and the American Cancer Society.

At the next visit, another crisis occurred. Mr. F had minimal urine output and the catheter was apparently clogged. Both Mr. and Mrs. F were distraught. When irrigation and an attempt to insert a new catheter failed, we agreed that Mr. F needed to go to the emergency room for reinsertion of the catheter. Mrs. F became very emotional and tearful and seemed to think that this situation had been her fault. My impression was that she felt that she had failed her husband by not having been able to prevent this. Mr. F was handling all this drama very well and appeared surprisingly calm and controlled. He tried constantly to console Mrs. F, but with little success. While we were waiting for transport, Mrs. F and I talked about the day's events and I managed to convince her to an extent that there was really nothing anyone could have done and that Mr. F would not be admitted as a patient and hospitalized because he was a hospice patient. This did seem to console her a little, and she calmed down and sat constantly by Mr. F's side until the doorbell rang. We agreed that they were to make contact with me when they returned home.

Comprehensibility = 2. The family's understanding of Mr. F's condition is such that Mrs. F is now competent enough to handle wound site dressings. Both Mr. and Mrs. F are very clear as to the correct positioning to relieve pressure from the wound site and have even devised their own method to that end. They are not able to transfer learning to new situations, however.

Meaningfulness = 2. The patient's attitude on life and death remains the same. He still aspires to return to his former physical condition, and his wife remains very supportive of his aspirations.

Manageability = 2. The patient remains very motivated, and the family is more resourceful. They have yet to realize, however, how well they are managing. The patient and the family are already managing to cope well with the stresses they have experienced. Through continued support from hospice, along with other community services, this situation should continue to strengthen.

Evaluation of the Plan: Rescoring on SSOC

Visits continued through the next 5 weeks, and the nurse worked with the patient and the wife both in their hope of recovery and their acceptance of death. The final scoring on situational sense of coherence at the time of the bereavement visit was as follows:

Comprehensibility = 3. The family had become completely in tune with the condition presented to them, and Mrs. F had become especially competent in all the tasks that were given to her. She was able to take on the responsibility of being the decision maker for needed care.

Meaningfulness = 3. Mrs. F had consistently shown that she was more than willing to work on problems as they developed and was a keen, intelligent, and capable associate. We had managed to work together as a team in our efforts to maintain Mr. F at an optimum level of comfort.

Manageability = 3. Mrs. F had been instrumental in managing her husband's pain. The idea of dying in pain had been one of his greatest fears. It was through her concerted and dedicated efforts that his decubitus ulcer had healed so beautifully and through her persistence and creativity that we were able to position him in such a way as to assist the healing process by relieving pressure from the site. Her extended family had eventually proved to be a pillar of support and had shown compassion and understanding when it was most required. The family had pulled together toward the end to manage the death in the manner that Mr. F had requested—he had remained at home to die.

Note

1. Adapted with permission from Beverly Brownell.

References

Aldous, J. (1990). Family development and the life course: Two perspectives on family change. *Journal of Marriage and Family, 52,* 571-583.

Allensworth, O., & Byrne, J. (1982, September). Sexism in the school: A hindrance to health. *Journal of School Health,* 417-421.

Antonovsky, A. (1987). *Unraveling the mystery of health: How people manage stress and stay well.* San Francisco: Jossey-Bass.

Bem, S. (1974). The measurement of psychological androgyny. *Journal of Consulting and Clinical Psychology, 42,* 155-162.

Boss, P., Doherty, W., LaRossa, R., Schumm, W., & Steinmetz, S. (Eds.). (1993). *Sourcebook of family theories and methods.* New York: Plenum.

Brewer, M., & Blum, M. (1979). Sex-role androgyny and patterns of causal attribution for academic achievement. *Sex Roles, 5,* 783-795.

Burr, W. (1973). *Theory construction and the sociology of the family* (pp. 199-217). New York: John Wiley.

Burr, W. (1991). Rethinking levels of abstraction in family systems theories. *Family Process, 30,* 435-452.

Chodorow, N. (1978). *The reproduction of mothering.* Los Angeles: University of California Press.

Curran, D. (1983). *Traits of a healthy family.* Minneapolis: Winston.

Duvall, E., & Miller, B. (1985). *Marriage and family development* (6th ed., pp. 3-64). New York: Harper & Row.

Feetham, S. (1991). *Feetham family functioning survey manual.* Washington, DC: Children's National Medical Center.

Feldman, L. (1982). Sex roles and family dynamics. In N. Walsh (Ed.), *Normal family processes* (pp. 354-379). New York: Guilford.

Fowler, J. (1980). Moral stages and the development of faith. In B. Munsey (Ed.), *Moral development, moral education and Kohlberg* (pp. 130-160). Birmingham, AL: Religious Education Press.

Fowler, J. (1981). *Stages of faith.* San Francisco: Harper & Row.

Fowler, J. (1984). *Becoming adult, becoming Christian.* San Francisco: Harper & Row.

Gilligan, C. (1982). *In a different voice.* Cambridge, MA: Harvard University Press.

Grusec, J., & Lytton, H. (1988). Sex differences and sex roles in social development. In E. Maccoby & N. Jacklin (Eds.), *Psychology of sex differences* (pp. 349-359). New York/ Berlin: Springer-Verlag.

Hansen, D., & Johnson, V. (1979). Rethinking family stress theory: Definitional aspects. In W. Burr, R. Hill, F. Nye, & I. Reiss (Eds.), *Contemporary theories about the family* (pp. 582-603). New York: Free Press.

Hayes, B., & Pittelkow, Y. (1993). Religious belief, transmission, and the family: An Australian study. *Journal of Marriage and Family, 55,* 755-766.

Hill, R. (1949). *Families under stress.* New York: Harper & Row.

Hill, R. (1986). Life cycle stages for types of single parent families: Of family development theory. *Family Relations, 35,* 19-29.

Hoffman, D., & Fidell, L. (1979). Characteristics of androgynous, undifferentiated, masculine, and feminine middle-class women. *Sex Roles, 5,* 765-781.

Janosik, E., & Green, E. (1992). *Family life: Process and practice* (pp. 3-66). Boston: Jones & Bartlett.

Jones, W., Chernovetz, M., & Hansson, R. (1978). The enigma of androgyny: Differential implications for males and females? *Journal of Consulting and Clinical Psychology, 46,* 298-313.

Kantor, D., & Lehr, W. (1975). *Inside the family.* San Francisco: Jossey-Bass.

Maccoby, E., & Jacklin, C. (1974). *The psychology of sex differences.* Stanford, CA: Stanford University Press.

Mattessich, P., & Hill, R. (1987). Life cycle and family development. In M. B. Sussman & S. Steinmetz (Eds.), *Handbook of marriage and family* (pp. 437-469). New York: Plenum.

McCubbin, M., & McCubbin, H. (1987). Family stress theory and assessment. In H. McCubbin & A. Thompson (Eds.), *Family assessment inventories for research and practice* (pp. 3-33). Madison: University of Wisconsin Center for Excellence in Family Studies.

McGoldrick, M., & Carter, E. (1982). The family life cycle. In F. Walsh (Ed.), *Normal family processes* (pp. 167-195). New York: Guilford.

Minuchin, S., Rosman, B., & Baker, L. (1978). *Psychosomatic families: Anorexia nervosa in context.* Cambridge, MA: Harvard University Press.

Nathason, C. (1977). Sex roles as variables in preventive health behavior. *Journal of Community Health, 3,* 142-155.

Patterson, J. M., & Garwick, A. W. (1994). Levels of meaning in family stress theory. *Family Process, 33,* 287-304.

Ransom, W., Schlesinger, S., & Derdeyn, A. (1979). A stepfamily in formation. *American Journal of Orthopsychiatry, 49,* 36-43.

Reiss, D. (1981). *The family's construction of reality.* Cambridge, MA: Harvard University Press.

Rodgers, R. (1962). *Improvements in the construction and analysis of family life cycle categories.* Unpublished doctoral dissertation, Western Michigan University, Kalamazoo.

Rodgers, R., & White, J. (1993). Family development theory. In P. Boss, W. Doherty, R. LaRossa, W. Schumm, & S. Steinmetz (Eds.), *Sourcebook of family theories and methods* (pp. 225-254). New York: Plenum.

Seligman, M. E. P. (1981). A learned helplessness point of view. In L. P. Rehm (Ed.), *Behavior therapy for depression: Present status and future directions.* New York: Academic Press.

Stokes, K. (Ed.). (1983). *Faith development in the adult life cycle.* New York: Sadlier.

Stuart, M. (1991). An analysis of the concept of family. In A. Whall & J. Fawcett (Eds.), *Family theory development in nursing: State of the science and art.* Philadelphia: F. A. Davis.

Watzlawick, P., Weakland, J. H., & Fisch, R. (1974). *Change: Principles of problem formation and problem resolution.* New York: Norton.

9

The Institution as Client

Institutional Interaction

Margaret M. Conger
Sandra G. Elkins

Intrasystem Assessment
Intersystem Interaction
Application of the Intersystem Model to an Institutional
    Problem
Conclusion

Because institutions are systems that are able to organize activities to carry out mutually agreed-on goals, a nurse is able to work with an institutional group as client using the same Intersystem Model principles that are used with the individual or family. Examples have already been given for use of the Intersystem Model with a client who is an individual or a family. In this chapter, the Intersystem Model will be applied to the institution as the client. The same assessment criteria using the intrasystem analysis of biological, psychosocial, and spiritual subsystems will be applied to the institution. The interaction between the client, the institutional group, and the nurse will be explored. The situational sense of coherence (SSOC) of an institution can also be examined, and interventions can be negotiated with an institution to improve its SSOC.

The application of the Intersystem Model is especially useful to an analysis of an institution because there are often major discrepancies between the fit of the institution and the value system of the nurse. The ability to examine the institution and how these value systems are operationalized can be an effective management tool.

A systems analysis of the organization based on the definition of Kuhn (1974) will be used in this chapter. Kuhn has defined an institution to be an organization in which two or more parties are engaged in some type of joint action to achieve a mutually agreed-on goal. To the extent that the organization acts as a unit, it is possible to analyze its behavior with respect to its detector, selector, and effector processes. Therefore, both an intrasystem analysis focusing on the organization itself and an intersystem analysis in which the relationship between the organization and the nurse is explored will be described. Much of what has been discussed in classic organizational theory would apply to the intrasystem analysis. The emphasis on behavioral management approaches that have surfaced in more recent times reflects intersystem interaction analysis.

Intrasystem Assessment

Using Kuhn's (1974) model, intrasystem analysis of the organization is concerned with how the detector, selector, and effector functions are accomplished. The nurse gains this information through analysis of each function: (a) *detector*—assessment of what is known about the interactional problem that arises from the subsystems of the organization, (b) *selector*—analysis of the value systems of the organization, and (c) *effector*—identification of typical behaviors and goals of the organization. Analysis of intersystem interaction is concerned with the patterns of communications and negotiations within the organizational context. These analyses are, of course, the basis of the Intersystem Model (Artinian, 1991) and will be the focus of this chapter. The goal of the entire process of assessing the information base of the organization, understanding the value system, and negotiating to make changes within the organization is to improve the SSOC of the organization. In so doing, the members of the organization will have a better understanding of the subtleties of the organization, learn to manage better, and demonstrate motivation to alter the organization's practices to better meet the organization's goals.

Using an Intersystem Model approach to the organization, the importance of the intersystem analysis cannot be stressed enough. An organization can meet its goals only through the people involved in the organization. As the values of both the organization itself and the people within the organization are explored, differences can be clarified. New ways to bridge differences in the value systems can be negotiated, leading to improvement in the output of the organization.

Detector Function:
Assessment of Developmental Environment

The information within the organization is examined during the analysis of the developmental environment of an organization. Inputs are experienced through the three subsystems identified in the Intersystem Model—the biological, psychosocial, and spiritual subsystems. Therefore, information about each of the subsystems will form the basis of the assessment because they provide the context within which an interactional problem develops.

Biological Subsystem Assessment

The biological subsystem of an organization itself is quite nebulous. Because the organization is composed of people, however, it is important to consider how the organizational structure affects the people within it. A number of health policies will affect the biological status of the members of the organization. Each organization will have health policies affecting sick leave time, health benefits such as wellness programs, and programs such as health insurance, which either enhance or hinder the wellness of the employees. Work schedules such as shift work, rotation of work schedules, vacations, and other such benefits will also affect the health of employees.

The predominant gender component needs to be assessed. If the workplace is primarily of one sex, some inspection of work policies that relate to that sex should be examined. Also, employee safety programs and environmental conditions that enhance the health of the employees need to be considered when doing a biological assessment of the organization. The actual structure of the building housing the organization can have an effect on the biological health of the individuals as well.

Psychosocial Subsystem Assessment

Because much of the interaction within an organization lies at the social level, a thorough assessment of the psychosocial climate of the organization is important. Inputs into the institution that will affect it are the management practices and values. Each organization has its own unique cultural climate, often expressed by a number of "unwritten" rules. These will affect the organizational climate. One cannot assess the organizational structure without paying attention to these rules. The cultural components that the employees bring to the organization are also important to the assessment process.

The presence of employee unions or other employee control processes needs to be assessed. Some organizations will have a committee structure in which employees within the organization will have a strong voice in determining institutional policies. In other institutions, such activities will be tightly controlled by upper management. The morale level of the people within the organization will be affected by these various management policies.

Spiritual Subsystem Assessment

When looking at health care organizations specifically, the spiritual subsystem must also be considered. Assessment of this subsystem would include determining if the institution has any religious affiliations. An institution representing a particular religious group will display many of the value systems of that group and will affect how decisions will be made within the organization. Even if the institution has no direct religious affiliation, the presence of a religious program such as a chaplain program will indicate something about the religious value system of the institution.

Selector Function: Analysis of Value Systems

Organizational Climate

The organizational climate is largely a reflection of the management style espoused by the leaders of the organization. One way to describe various management styles is in terms of the focus of the leader as being task oriented or people oriented. Recent work on leadership styles is suggesting a third management style of situational leadership.

Task-oriented leadership can also be described as bureaucratic or pyramidal. In this type of leadership, the power is centered at the top, with the emphasis on accomplishing a specific task. The worker is viewed as a means to an end, with very little intrinsic value to the final product. Taylor (1911) was one of the earliest theorists in leadership. His focus was improvement of productivity through improvement of techniques. A similar viewpoint was expressed by Gilbreth (1973), who made time and motion studies famous. McGregor (1960) described this type of organizational leadership as Theory X, in which the worker was viewed as unmotivated and in need of close supervision. The manager's role is that of determining the structure of the work setting and maintaining close control over workers' actions.

The leadership situation in many hospitals continues to be very hierarchical. Decisions affecting workers are still often made by leaders who are not directly involved in the implementation of the decision. The following vignette[1] told by Fran (pseudonym) illustrates these differences. Fran had a strong commitment to patient education and had worked hard to establish a cardiac rehabilitation program.

Because we have a staffing shortage right now, there are times when I have to cancel a class. I feel that we should be able to provide the clients with a service that we promise them, but sometimes we can't.

Fran went to her manager, who was a physician in charge of the cardio-respiratory program. He did not see the urgency of the teaching program and thus refused to support Fran in her attempt to get better staffing so that the classes could be held as scheduled. His focus was on maintaining a profitable department. Fran was unable to negotiate for the staff necessary to maintain the education program.

A more human-centered approach to management has been developed. Mayo (1945) stated that, in addition to finding good technological solutions to increasing productivity, attention also had to be paid to the workers themselves. His work at the Hawthorne Works of the Western Electric plant demonstrated that when workers were given special attention, their productivity increased. Consideration for the feelings and attitudes of the workers was central to good management. McGregor (1960) described the human relationship approach to leadership as Theory Y, in which workers were considered to be self-directed. In addition, self-control in the worker is considered to be indispensable to accomplishing the organizational goal. This concept was further described by Argyris (1971). Herzberg (1966) identified a number of factors that were motivational for workers. All these theorists have emphasized the need for establishing a work environment in which the worker is able to have considerable control over the environment.

Recently, the concept of situation leadership has been promoted. In this approach, the effective leader is able to alter his or her leadership style to fit the needs of the worker. Emphasis is placed on identifying worker behavior and then utilizing an appropriate management style. Thus, the effective manager will be one who is able to alter leadership style based on the needs of the worker. When working with a person with low motivation, emphasis on a task-centered approach will provide the most effective outcome. When managing a highly motivated worker, however, a more person-centered approach will be most effective. Theorists who have been prominent in this movement include Tannenbaum and Schmidt (1958), Fiedler (1967), House and Mitchell (1974), Stinson and Johnson (1975), Vroom and Yetten (1973), and Hersey and Blanchard (1974).

The maturity of the worker has been identified as an important consideration for any manger to understand. Argyris (1962) suggests that in many organizations there is a collective effort to keep workers from becoming mature; the workers are encouraged to be passive, dependent, and subordinate. The organization thus retains control over how the organizational goals will be met.

This behavior is not uncommon among nursing managers. One young nurse, Sarah, reports on two situations in which she was assigned to care for patients with multiple complications. After the second experience, she stated "I couldn't take care of anybody too difficult right now. I need to take easier patients. They (nursing managers) realize that I am not ready to handle difficult cases." Sarah's nurse manager supported her in this behavior and allowed her to remain dependent on other staff nurses to care for those more difficult patients. The nurse manager did not see this situation as one in which Sarah could have been encouraged to grow.

Organizational Cultural Assessment

Organizational culture has been described as the pattern of basic assumptions and shared values that are developed within a group to aid in its survival (Schein, 1985). Schein suggests that there are three levels of organizational culture: (a) the physical environment, (b) the values held by members of the organization, and (c) the basic underlying assumptions held by the members of the organization. The values and underlying assumptions of the organization result in expectations for group member behaviors. These expectations form a means of internal control over the participants. Such expectations have generally risen out of a need to solve work-related problems and to allow the group to survive in the workplace (Coeling & Simms, 1993).

These behaviors form a very basic value system among the members. As new members come into the group, they are quickly socialized into the group norms and adapt their behaviors to fit those of the group if they are going to be successful in that group. The group serves as a "reference group" (Shibutani, 1994) for new members and is used by the newcomers to make comparisons about themselves. Research has shown that new members will often adopt behaviors of the group that are quite inconsistent with their former behavior. Thus, it can be seen that group expectations for behavior can be extremely strong.

Not all nurses follow the group culture, however. Although they take considerable risk in deviating from the culture of their workplace, their perception of their role as a nurse can motivate them to take an alternate path. Lucy's story illustrates how a nurse may deviate from the group cultural norm:

A baby (born very prematurely) was brought to the nursery to die because they (the doctors) felt it was too young to survive. The baby had a heartbeat and was breathing, but its chance for survival at that time, which was in the late seventies, was not as great as it would be today. I did what I thought was comforting to this baby and followed orders, so I did not start oxygen and did not give any medications or anything like that and tenderly cared for this baby.

The practice of the nursing staff in this unit was to wrap the baby in a blanket, place it in an isolette, and ignore it. The other nurses believed that this practice assisted the mother in accepting the death of her child.

Lucy, however, was not comfortable with this cultural norm. By viewing the situation from the perspective of the mother and baby and taking into account their values, she went against the usual unit procedure and attended to the baby and mother with great sensitivity. Eventually, she was able to develop her own practice in a more holistic manner and served as a change agent to alter practice in neonatal nursing.

The strength of the group culture can be a formidable hindrance to effecting change. If a manager proposes to make changes that are in conflict with the organizational culture, extensive control measures will be needed to effect the change. Activities such as close supervision, rewards and punishment systems, and other control measures will be needed. When there is a fit between the activities imposed by the leader and those supported by the group culture, change can occur readily. Thus, when a manager needs to make an organizational change, strategies to influence the culture of the work group will prove to be more effective than control strategies.

Organizational culture focuses on the work life of the members as opposed to focusing on every aspect of living. Members of the organization come to the workplace with their own religious, ethnic, and family cultural norms. If these are in opposition to the organizational culture, conflict may ensue.

In addition to differences in the personal culture that each member of the group brings to the workplace, differences in cultural values can exist between groups within the organization. Hospitals have been described as having three distinct cultural values espoused by members of the organization: The management espouses a business culture, the physicians espouse a cure culture, and the nursing staff espouses a care culture. The hospital manager often comes out of a business culture in which profit and loss, efficiency of workers, and other business values are stressed. The values nurses bring to the hospital are quite different. Values such as caring for the individual person requiring nursing care, advocacy for this client, and a concern for open access to health care for all persons have been well documented (Cooper, 1991; Gadow, 1985; Leininger, 1984; Peter & Gallop, 1994; Watson, 1988).

In addition to the management and staff groups common to the typical organizational structure, hospitals have an additional group, the physicians. Although physicians do not work directly for the hospital—except in the health maintenance organization structure—the hospital is still viewed by physicians as their workshop. This is the place to which physicians will bring patients in need of specialized health care services. The addition of a third group in the hospital organizational structure increases the possibility for dissension.

Although these organizational cultural differences exist between physicians and nurses, it is still important for these groups to work together cooperatively to achieve positive patient outcomes. Historically, Stein (1967) described the interaction as the doctor-nurse game, with the physician superior to the nurse in a hierarchical order. The nurse role was to make recommendations concerning the patient's welfare in a manner that made it look as though the suggestion came from the physician.

This relationship is no longer considered to be in the best interest of the patient. Stein, Watts, and Howell (1990) suggest that the rules of the game are changing. Nurses are increasingly becoming specialists in their chosen area and are seeking increased independence and responsibility. As the nurse becomes more independent in practice, conflict between the doctor and nurse is a common outcome. The following vignette illustrates this conflict.

Greg was working as the night supervisor when this incident occurred. He had been called to the cardiac step-down unit to see a patient who had just experienced an episode of complete heart block. Greg stated,

> I called the physician to get an order to move the patient to ICU for a pacemaker insertion. The physician started screaming at me and said that I could not transfer his patient. He recently had a fight with the director of ICU and didn't want any of his patients to be placed in ICU.
>
> Well, you are taught to follow the requests of the physician, but you are also taught to be a patient advocate. I knew that there was no way this patient was going to stay in the step-down unit.
>
> What I did was to say to the physician, "I'm sorry, but we're going downstairs to ICU now. I have the resident with me and we are taking the patient there whether you write the order or I write it for you."
>
> He started screaming at me, "I'll have your job." I said, "You can have this job but we will talk about this in the morning."
>
> Well, I got the patient down to ICU and got him set up with a temporary pacemaker. The patient did well, throughout the rest of the night.

Greg demonstrated patient advocacy and assumed a measure of risk when he wrote a nursing order for the transfer of the patient to the ICU. He represented the patient's best interests; he had changed the rules in the doctor-nurse game.

The cultural value held by many nurses to provide quality of care to all patients may come in conflict with the staffing patterns of the institution. One nurse recounts her experience with this problem:

> I have always enjoyed oncology. I know this sounds strange, but I enjoyed it because I liked the idea of being able to give these patients who were dying the support that they needed. But there was never any time available to do that. You went in and gave

them their medications; you gave the chemotherapy and ran out of the room. You
didn't really have time to answer their questions, to give them any comfort. To me,
that is one of the real big reasons that I went into nursing.

This nurse felt so compromised in her ability to care for patients in a manner
consistent with her value system that she asked for a transfer to a nursing unit
where the acuity of the patients was less and she would not have to face this
predicament on a daily basis.

At times, the personal culture of a staff member may conflict with the
institutional norm. For example, the staff nurse with a strong religious
conviction on the taking of a life may strongly object to assisting with an
abortion. If abortions are a common practice in the hospital in which she
works, however, the nurse will be in conflict with the manager's expectations
for all staff to accept nursing assignments where most needed.

These personal cultural differences can cause some confusion in the nurse;
they can also serve as a stimulus for growth. One nurse expressed this growth
in the following way:

From my growing up and all that stuff you are taught . . . from the moment of
conception that you are a human being. However, there are times when you have to
weigh the mother's life against the fetus's life and it really brings a challenge to what
you are thinking about. You want both to survive, but you know that one can't and
one will.

Effector Function:
Identification of Organizational Goals

For an institution to thrive, it is necessary for it to be aware of and find ways
to meet its organizational goals. For some institutions, empowerment of the
workers is a key concept in achieving its mission and goals (Tebbitt, 1993).
This empowerment arises from a paradigm shift from organization control to
that of personal partnership and participation of all workers in decision-
making functions. The use of the Intersystem Model in working through the
intersystem relationships between the organization and the staff is an effective
way of accomplishing this paradigm shift. An evaluation of the organizational
SSOC as it relates to a specific problem is a starting point in making such an
assessment. Because most hospitals still remain in a hierarchal organizational
structure, much of the communication of information is from the top down.
Recent endeavors have been made to alter this structure.

In nursing, the current movement is to decentralize management so that
decisions are made closer to the people who will be affected by them.
Providing all staff members with a means to communicate their concerns and

to have a measure of self-governance has been shown to increase the motivation of workers to support the organizational goals. The development of professional governance models has done much to promote this type of communication. Jenkins (1991) describes a professional governance model as one in which all members of the organization give input into the decision-making process. The person who will be responsible for carrying out a function is involved in the decision-making process. For example, the staff nurses have control of decisions that directly affect patient care, the unit manager has control of decisions concerning how the unit is run, and the staff educator makes decisions about educational activities. In an organizational setting such as this, the organization chart appears to be more of a multiarmed centipede rather than the typical ladder approach of the hierarchal structure.

Intersystem Interaction

In the organizational structure that has been typical of hospitals, the information flow is from the top administrator down to the workers. In such a setting, the nurse manager must bridge the gap between two cultural groups: the management group and the staff group. The needs of both subsystems must be identified and communicated to allow for communication between these groups. The challenge for the manager is twofold: (a) to find creative ways to shape the change proposed by the upper management level to fit the culture of the staff nurse group and (b) to communicate the needs of the staff group to the management group. If the gap between the cultural norm of the staff group and the proposed change is large, considerable external force will be required to effect the change (Coeling & Simms, 1993). The cultural gap needs to be identified and a means to lessen it found for successful change to occur.

The focal goal of adaptation within the institution involves the following activities: (a) communication of information, (b) negotiation of values, and (c) development of management policies that promote the motivation of the workers to support the organizational structure. Each of these affects the others. The organizational structure will dictate how information can be communicated formally, what channels are available for negotiation, and what incentives can be offered to encourage participation in the change process.

Communication of Information

An intersystem assessment necessary to every organization is that of the communication system. All organizations use both formal and informal communication mechanisms to share information. The formal flow is often defined in an organizational chart depicting how information is supposed to be transmitted. The downward mechanisms often consist of policy manuals, memos, computer message systems, or group meetings. The upward flow mechanisms include suggestions systems, department meetings, committee reports, or "open door" policies of managers. The informal flow of information is equally valuable to an organization. This flow often bypasses the mechanisms established by the managers and can be quite effective. The information may move upward, downward, or laterally and skip management levels. When assessing an organization, all these means of communication are important to assess. It is through communication mechanisms that the purposes of the organization can be shared.

Negotiation of Values

The art of negotiation is necessary for all workers in the organizational setting because of conflicting needs, wants, and desires of the members of the group. The values of the employee are often not in synch with organizational values. When this situation arises, the employee will have to either adapt to the existing organizational culture or develop ways to renegotiate with managers or other employees to find resolution to the conflict. Smeltzer (1991) has described negotiation to be either cooperative or competitive. In cooperative negotiation, everyone wins; in competitive negotiation, only one party wins. To retain harmonious work relationships, cooperative negotiation will be far more effective.

The successful negotiator prepares well prior to initiating the process. She or he will gather as much knowledge as possible prior to the session. Marquis and Huston (1994) emphasize that the negotiator must know both the upper limit and the lower limit of the expectations of the situation. They suggest that it is best to start with the upper limit, realizing that in the negotiation process some give will be necessary. One should not disclose what the lower limit is, however, because in doing so, one's power base will be eroded. During the process, the negotiator should introduce the topic, encourage discussion during the conflict stage so that the values of all participants are taken into consideration, prepare a compromise position that summarizes the various points of view, and finally, when agreement is reached, recap it so that everyone has a clear understanding of what will happen next.

The term *cultural brokering* has been suggested to describe the type of negotiation that occurs when one needs to bridge a gap between two cultural groups (Jezewski, 1993). It is effective in situations in which the two groups are of unequal status, such as the staff nurse and the hospital management or staff nurse and physician. In such situations, it is necessary to know how to circumvent the barriers that are present in preventing the group with less power from being heard. Three aspects of brokering are needed: (a) the ability to mediate when the two groups are in conflict, (b) the ability to negotiate to reach an agreement, and (c) the ability to be innovative when new solutions to the problem are required.

The experience of Amy illustrates how nurses can broker successfully despite their less powerful position in the hospital hierarchy. Amy frequently admitted elderly patients from a nursing home to her medical unit who had a "Do Not Resuscitate" (DNR) order at their residence. On admission to the acute care hospital, however, these orders were ignored by a particular physician.

Amy found that this physician refused to initiate discussion of a DNR status with the patient's family. She said, "Whenever we approached this doctor, he would blow you off and would not deal with it." Despite the powerless position that Amy found herself in, she was able to find an alternate approach that proved to be effective. She said, "If you have the opportunity to meet the family, you can discuss it with them and then have them talk to the doctor directly."

When the physician is confronted directly by the family, he or she can no longer ignore the issue. This vignette provides an excellent example of how it is possible to negotiate from a less powerful position.

The art of negotiation is an important means to bridge the gap between the value systems of the management and the workers. Time spent identifying the values of each before attempting to make change is well spent. Clear communication so that all know the issues under consideration is important. Finally, looking at the resources that the workers have to manage the change is important. Using the construct of the SSOC to score the institution on these behaviors will provide the manager with a clearer understanding of the total situation.

Organizational Ethics

An important negotiation process that must occur in every organization is that of negotiating on ethical issues. Trunfio, Auday, and Reid (1992) suggest that a moral corporate culture promotes mutual understanding and responsibilities that result in virtues such as "veracity, wisdom, justice, temperance, and courage" (p. 23). All institutions face ethical considerations concerning interactions both with workers and with their public. Hospitals face an increased ethical responsibility because of their service to the public in situations that are often life threatening. The values of the institution will have some bearing on how each issue is evaluated.

Ethical decisions are never made easily. White (1991) states that

> no matter what you do in the case of an ethical dilemma, there is no satisfactory answer or solution. Each situation has unique considerations that must be evaluated

on [their] own merits. Often the ethically correct action is not economically expedient. (p. 21)

Values concerning life, the definition of person, and the religious beliefs of the institution will affect how the institution reacts to situations. Because this is such an important consideration in health care institutions, many have established ethics committees to work through this decision-making process.

Hospital Ethics Committees

The rise of hospital ethics committees has occurred in response to the increasing number of difficult ethical issues that are present in the very high technological environment of modern health care. Using a mix of medical expertise, practical wisdom, and careful deliberation, decisions can be made that reflect more than the limited perspective of a wholly private decision-making process (Murphy, 1989). The decision is also better made in this setting than in the more impersonal court setting. To make a sound ethical decision, one needs adequate information, multiple perspectives, deliberate thought, emotional support from all decision makers, and legal acceptability (Murphy, 1989). The involvement of an array of disciplines on the ethics committee, including representation from the medical staff, nursing staff, chaplains, social workers, and lay people, is important.

Organization of Behavior: Case Example of an Ethical Decision-Making Process

What follows is a recent example of how the nature of the institution affects the ethical decision-making process.

A set of twins conjoined, sharing a liver and heart, was born at Loyola University Medical Center, a Catholic institution. After months of agonizing discussions and study, the hospital staff made the decision to encourage the parents to keep the twins "warm, fed, and cuddled until they died" (Brandon, 1994a). The parents did not accept this opinion and had the twins transferred to Children's Hospital in Philadelphia for possible surgery. There, a decision was made to proceed with the surgery, knowing that one of the twins would die as the result and the other had a very marginal chance for survival. The ethical decision made at this hospital was "If somebody thinks he can save the life of one baby, shouldn't he try?" (Brandon, 1994b).

Each of these institutions viewed this situation differently. The Catholic hospital staff focused on values in which extraordinary means to sustain life were not considered necessary. On the other hand, the Children's Hospital staff focused on the technical capabilities available in its institution and their belief that a surgical intervention could possibly be successful. Both belief systems were congruent with their institutional value systems, but it is the value system of the parents that must guide the decision-making process.

Application of the Intersystem Model to an Institutional Problem

To illustrate how the Intersystem Model can be applied in an institutional setting, the following case study developed by Gigi Gaughran, a graduate student at Azusa Pacific University, is presented.[2] In this example, the nurse manager is in the role of the change agent. The staff nurses represent the institutional group. The case study illustrates how the Intersystem Model can be used to bring about effective change using the organizational theories described previously.

Focus of the Situational Interaction

The p.m. staff of a rehabilitation unit approached the nurse manager with the complaint that when the patient census was over 10, it was too much work for them to shower all the patients in the evening.

Assessment

Developmental Environment

Biological Subsystem

All the staff nurses were female and physically able to perform the strenuous nursing care activities required by rehabilitation patients.

Psychosocial Subsystem

The p.m. staff was stable, quite cohesive, and able to communicate with each other well. The three RNs on the unit each had different communication styles but were effective workers. Two of the three nursing technicians were hardworking and cooperative. The third would do only what was required and would never volunteer to assist others. Some negative attributes identified through the intrasystem assessment included staff difficulty in seeing options for change because they had been working there so long. There was also some unresolved hostility toward the nursing technician whom the group considered lazy. Finally, there was an "us" versus "them" attitude between the day and p.m. shift nurses.

Spiritual Subsystem

Nothing is known about the spiritual orientation of the group.

Situational Environment

Intrasystem Analysis of the Group

Detector Function. The p.m. nurses knew that downsizing of the unit had occurred and that no new staff could be hired to meet additional patient needs.

Selector Function. The p.m. nurses valued what they did and believed their jobs had a positive impact on the lives of their clients.

Effector Function. The group responded to the complaints of one RN and worked into a frenzy of complaints about their heavy workload and their inability to provide adequate care.

Intrasystem Analysis of the Nurse Manager

Detector Function. The manager believed she knew her staff well and had their support despite some unpopular decisions she had made based on the need to downsize the unit.

Selector Function. The manager valued a good working relationship with the staff and valued a people-oriented leadership style.

Effector Function. In the year she had been unit manager, she had demonstrated her ability to effect change. Therefore, she developed a plan to encourage negotiation between the staff groups to resolve this problem.

Analysis

Identification of Stressors

Recent downsizing of the unit

Hostility between p.m. and day shifts

Coping Resources

Stable staff with good communication skills

Track record of manager in bringing about effective change

Scoring on Situational Sense of Coherence

Manageability = 1. The staff felt they had no control over the situation.

Comprehensibility = 2. Although they understood why the decision had been made to give showers on the p.m. shift, they still felt that they were being "dumped on" by the day shift.

Meaningfulness = 3. The staff felt the problem was of sufficient importance to them to put time and effort into a solution.

Nursing Diagnosis

Impaired social interactions related to conflict regarding task assignment.

Nursing Goals

1. To assist staff to develop a plan to resolve problem
2. To increase cooperation among the shifts

Intersystem Communication of Information and Negotiation

The nurse manager felt that negotiation between the day and p.m. staff nurses would be more appropriate than issuing a mandate for the solution. To prepare for the negotiation, the manager identified the problem to the staff and set up some guidelines for how the negotiation would occur. The nurse manager realized that the p.m. shift believed that they had much to gain from negotiation, whereas the day staff believed they only had something to lose. Such a situation could well end in a stalemate. Thus, she prepared a contingency plan in case the first meeting did not produce a satisfactory outcome.

The negotiation meeting was planned to last for only 55 minutes. During this time, each staff member was encouraged to present ideas to solve the problem. Although the meeting started off with much positive input, it soon deteriorated into a controversy between the two shifts. Because no consensus could be reached, the nurse manager stopped the meeting and told the staff that they would reconvene in 1 week. In addition, she told them that if they could not develop a plan, she would have to initiate one herself.

During the week between meetings, the staff had the list of suggestions that had been presented at the first meeting. These stimulated considerable conversation. At the second meeting, the staff displayed a more cooperative spirit. They felt that they all had something to gain by deciding on which of the options presented would work best for both shifts because now they understood that if they could not develop a plan, the nurse manager would impose her own plan.

Organization of Behaviors: Mutual Plan

The final decision was to include the patients in the decision process. On admission, each would be asked if he or she preferred a morning or evening shower. The day and night shift nurses would help with showers of those patients who preferred the morning shower. Also, there would be some redistribution of hours between the nursing technicians so that more staff would be available during the early hours of the p.m. shift.

Evaluation of the Plan: Rescoring on SSOC

The nurse manager's evaluation of this process was that the staff believed that they had input into a work situation of importance to them and were able to come to a satisfactory solution. On reevaluation of the staff on their SSOC, the scores were improved. On comprehensibility, she rated them as a 3 because the p.m. shift no longer felt that they were being "dumped on." On meaningfulness, she rated them as a 3 because both shifts had made a strong commitment to make the new plan work. On manageability, she rated them as a 2 because, although the staff believed that there were many viable options open to them, they were not quite sure that staffing was adequate to make the agreed-on plan really work. To overcome this lack of trust, the plan was to be reevaluated in 3 months.

This case study illustrates how a nurse using the Intersystem Model can solve problems in the institutional setting. The ability to assess the values of both subsystems, the institutional group, and the manager is an asset when working in organizational settings. Finally, the use of SSOC provides an objective way to evaluate the effectiveness of changes in the organizational structure.

Conclusion

In this chapter, theories related to the interaction between the nurse and the institution have been described. It illustrates how the Intersystem Model can be applied in this broader setting. As nurses learn to evaluate both the value system of the institution in which they work and their own value system, they can be more effective in negotiating for changes that will allow the institution to achieve its goals. At the same time, interaction will provide for an environment in which each worker will be able to increase his or her maturity level. The interpersonal interaction between the institution and worker can have positive rewards for both.

Note

1. Although each of the vignettes presented in this chapter are actual experiences reported to one of the authors, the names of the nurses have been changed to maintain their confidentiality.
2. Adapted with permission from Gigi Gaughran.

References

Argyris, C. (1962). *Interpersonal competence and organizational effectiveness.* Homewood, IL: Irwin, Dorsey.

Argyris, C. (1971). *Management and organizational development: The path from XA to YB.* New York: McGraw-Hill.

Artinian, B. (1991). The development of the Intersystem Model. *Journal of Advanced Nursing, 16,* 194-205.

Brandon, K. (1994a, February 20). More than just life—or death—decision. *Chicago Tribune,* pp. A1, A20.

Brandon K. (1994b, February 21). For survivor, "What are we really creating?" *Chicago Tribune,* pp. A1, A6.

Coeling, H. V. E., & Simms, L. M. (1993). Facilitating innovation at the nursing unit level through cultural assessment, Part I. *Journal of Nursing Administration, 23,* 46-53.

Cooper, M. C. (1991). Principle-oriented ethics and the ethic of care: A creative tension. *Advances in Nursing Science, 14,* 22-31.

Fiedler, F. (1967). *A theory of leadership effectiveness.* New York: McGraw-Hill.

Gadow, S. (1985). Nurse and patient: The caring relationship. In A. H. Bishop & J. R. Scudder (Eds.), *Caring, curing, coping* (pp. 31-43). Tuscaloosa: University of Alabama Press.

Gilbreth, F. (1973). Applied motion study: A collection of papers on the efficient method to industrial preparedness. In F. B. Gilbreth & L. M. Gilbreth (Eds.), *Hive management history series no. 28.* Easton, PA: Hive.

Hersey, P., & Blanchard, K. H. (1974, February). So you want to know your leadership style? *Training and Development Journal,* pp. 1-6.

Herzberg, F. (1966). *Work and the nature of man.* New York: World Publishing.

House, R. J., & Mitchell, T. R. (1974, Autumn). Path-Goal theory of leadership. *Journal of Contemporary Business,* p. 81.

Jenkins, J. (1991). Professional governance: The missing link. *Nursing Management, 22*(8), 26-30.

Jezewski, M. A. (1993). Culture brokering as a model for advocacy. *Nursing & Health Care, 14,* 78-85.

Leininger, M. M. (1984). *Care: The essence of nursing and health.* Thorofare, NJ: Slack.

Kuhn, A. (1974). *The logic of social systems: A unified, deductive, system-based approach to social science.* San Francisco: Jossey-Bass.

Marquis, B., & Huston, C. J. (1994). Negotiation. In B. J. Marquis & C. J. Huston (Eds.), *Management decision making for nurses* (pp. 302-315). Philadelphia: J. B. Lippincott.

Mayo, E. (1945). *The social problems of an industrial civilization.* Boston: Harvard Business School.

McGregor, D. (1960). *The human side of enterprise.* New York: McGraw-Hill.

Murphy, P. (1989). The role of the nurse on hospital ethics committees. *Nursing Clinics of North America, 24*(2), 551-556.

Peter, E., & Gallop, R. (1994). The ethic of care: A comparison of nursing and medical students. *Image: Journal of Nursing Scholarship, 26,* 47-51.

Schein, E. H. (1985). *Organizational culture and leadership.* San Francisco: Jossey-Bass.

Shibutani, T. (1994). Reference groups as perspectives. In N. J. Herman & L. T. Reynolds (Eds.), *Symbolic interaction: An introduction to social psychology.* Dix Hills, NY: General Hall.

Smeltzer, C. H. (1991). The act of negotiation: An everyday experience. *Journal of Nursing Administration, 21*(7/8), 26-30.

Stein, L. I. (1967). The doctor-nurse game. *Archives of General Psychiatry, 16,* 699-703.

Stein, L. I., Watts, D. T., & Howell, T. (1990). The doctor-nurse game revised. *New England Journal of Medicine, 322,* 546-549.

Stinson, J. E., & Johnson, T. W. (1975). The Path-Goal theory of leadership: A partial test and suggested refinement. *Academy of Management Journal, 18*(2), 242-252.

Tannenbaum, R., & Schmidt, W. H. (1958, March/April). How to choose a leadership pattern. *Harvard Business Review,* pp. 95-102.

Taylor, F. W. (1911). *The principles of scientific management.* New York: Harper & Brothers.

Tebbitt, B. V. (1993). Demystifying organizational empowerment. *Journal of Nursing Administration, 23*(1), 18.

Trunfio, E. J., Auday, B. C., & Reid, M. A. (Eds.). (1992). *Developing moral corporate cultures.* Wenham, MA: Gordon Institute for Applied Ethics.

Vroom, V. H., & Yetten, P. W. (1973). *Leadership and decision making.* Pittsburgh, PA: University of Pittsburgh Press.

Watson, J. (1988). Human caring as a moral context for nursing education. *Nursing and Health Care, 9,* 423-425.

White, G. (1991, October). Students, too face ethical issues. *The American Nurse,* p. 21.

10

The Community as Client

Involving Community

Barbara M. Artinian

Theoretical Models of Community Participation
Nursing Leadership to Promote Community Participation
Community Assessment
Involving Community: Case Analysis of a Healthy Cities
 Project
Clinical Example
Discussion

The goal of the community health nurse is to promote and preserve the health of aggregates, communities, or populations through organized community activity designed to change lifestyle patterns, implement new social policies, and improve the environment. To do this, attention must be paid to the interpersonal interactions taking place within the community and between the care provider and community members. In addition, all the structural constraints impinging on the community, such as economic, religious, political, and legal factors, must be considered. The Intersystem Model provides a format for assessing and working with the community that takes into account community definitions and knowledge of health, community values regarding health, and community behaviors carried out to promote health. It also provides direction for acquiring population-focused data necessary to identify community health problems and develop mutual plans for resolving them.

In this chapter, four aspects of community participation are discussed: (a) theoretical models for organizing community participation, (b) leadership styles needed for coordinating community participation, (c) a plan for assessing community data and participation in health care derived from the Project GENESIS program and the Intersystem Model, and (d) a case example illustrating how community participation could be used to improve health. Using the Intersystem Model framework, a secondary analysis of data from the doctoral dissertation of Jean Yan (1993) is presented as the case example. I served as a committee member for the dissertation and provided guidance for the grounded theory analysis of community participation.

Community is broadly defined by *Merriam-Webster's Collegiate Dictionary* (1994) as "the people with common interests living in a particular area" or as "an interacting population of various kinds of individuals in a common location" or "a group of people with a common characteristic or interest living together within a larger society." These definitions encompass both the terms *aggregate* (interacting persons with a common purpose or purposes) and *community and population* (aggregate of people sharing space over time within a social system) as defined by Swanson and Albrecht (1993, p. 8). A simple definition of community is given by Yan (1993): "a group of people living in a defined area sharing similar values and interests" (p. 34). Community involvement is described by Tsouros and Bracht (1990) as the process by which the community assumes its responsibility for its own welfare and develops the capacity to contribute to the residents' as well as the community's development.

In using the Intersystem Model to plan care for communities, just as when using it to plan care for individuals or families, the nurse must be aware of the power differences between the health care professional and the clients he or she is working with. For a true partnership to occur, the interactional process must be carefully monitored to be sure that input from both groups is honored. The problem of inequality of power is further complicated in the community setting because the socioeconomic, political, and legal bases for problems are more evident and the clients often do not have the resources or skills to challenge the system. Although community participation is a goal, many health planners are uncertain about how to achieve it. Often, goals for community participation are vague and unrealistic or simply ignored, with the result that planners act from their own perspectives with no input from the community. De Kadt (1982) has commented on the reasons for lack of community participation: "The concept of community participation has popularity without clarity and is the subject of faddishness and a lot of lip service" (p. 174). The case study presented in this chapter is an example of this. A careful analysis of the interaction between the planners and community members using the Intersystem Model framework illustrates the difficulty of achieving true mutuality in organization of behaviors and the negative effect on outcomes when mutuality does not occur.

A number of very useful case examples are given in *Community Health Nursing: Promoting the Health of Aggregates,* edited by Swanson and Albrecht (1993), to illustrate how nurses can identify community health problems and then work with individuals, families, and whole community groups to mutually resolve them. For example, the problem of lead poisoning was addressed through forming a community coalition group to work with the professional groups (pp. 587-592). Not enough detail is given about the interactional process that took place in the various examples, however, to understand how the power discrepancies among the interactants were handled. Again, the use of the Intersystem Model format would make clear how values were negotiated and plans were organized.

Theoretical Models of Community Participation

Community health nurses work within communities in cities or rural areas to organize community effort to make the area a more healthy place to live. A healthy city is defined by Hancock and Duhl (1988) as one that is "continually creating and improving those physical and social environments and expanding those community resources which enable people to mutually support each other in performing all the functions of life and in developing to its maximum potential" (p. 6). It requires a commitment to health and an infrastructure to support its activities. It also requires a commitment of both citizen and official groups to participate in mutual planning and decision making.

Community participation has been identified as an important characteristic of healthy communities. Hickman (1993) has summarized the work of Cottrell (1976) and Goeppinger, Lassiter, and Wilcox (1982) to define aspects of the healthy community. They are commitment, participation, articulateness, effective communication, conflict containment and accommodation, self-awareness and other awareness, clarity of situational definitions, and management of relations with larger society (p. 133). These are characteristics of a mature community that can participate in the planning and problem solving at all levels of a change process and take ownership of the change. Change introduced through partnerships in planning will have long-term effects, in contrast to rapid change that is imposed on a community without its input.

The World Health Organization (WHO) has developed the Healthy Cities Initiative to promote health in cities. The keystone of the WHO program is community participation because community members are thought to be best able to represent their health needs and values. Yan (1993) identifies the benefits of community participation for citizens: (a) It heightens the community's awareness of its own capabilities to make choices and to influence outcomes, (b) it strengthens interpersonal relationships and fosters self-confidence, (c) it reduces feeling of powerlessness and alienation, and (d) it improves socioeconomic conditions (p. 5).

From her review of the community participation literature, Yan (1993) identified three typologies of community participation: the development model, the market model, and the democratic model.

Development Model

The focus of the development model is on community involvement, control, development, and collaboration. Deliberate strategies are used to mobilize communities such as encouraging social action, self-help and grassroots organizations, and community planning. The purpose is to heighten the capabilities of the citizens to make choices and to influence outcomes. This is often done by making people aware of their life situation, why it is so, and what alternatives they have to address the problems they identify.

Market Model

The focus of the market model is on efficiency of public service delivery. It considers consumer participation, coproduction, and coprovision. The assumption is that public services are joint products of the activities of both consumers and public service agents. Both groups determine what public goods and services should be supplied and in what quantities. For public choice to be effective, Ostrom and Tiebout (1961) argue that units of government should be smaller so that they can respond to changes in citizens' needs and preferences over time. The goal is improved service delivery and distribution. Processes such as voucher systems, resource pooling, and time or money donations are used.

Democratic Model

The democratic model focuses on citizen involvement and public action through building coalitions and changing regulatory practices. It is based on the premise that citizens have the right to participate in and influence decision making that affects them. Citizens are expected to share equally in the benefits and burdens of public policies. The democratic model is a way of involving citizens in the decision-making process and sensitizing those in power to the multidimensional aspects of their constituencies and the impact of their decisions. Systems theory provides the framework for assessing the relationship between citizens and bureaucratic organizations. The more open the organizations are to feedback from the community, the more adaptive and responsive to the changing needs of the community they will be. It describes a situation in which organizations recognize that the people have knowledge of community problems that may not be understood by outsiders and then use what is learned to coordinate community programs. By planning with the community and linking knowledge with action, a "critical fit" can be achieved between the ability of the community to define and communicate its needs and the programs that are developed to meet those needs.

Intersystem Model

In each of the community participation models described previously, community participation is encouraged, but there is no provision made for the inequality of power between the citizen groups and the official groups. Therefore, none appears to meet the objective of the Healthy Cities Project that city dwellers be empowered to effect change in health by improving their opportunities to fully participate in committees constituted to implement health care programs. Therefore, they would not provide guidance for the community health nurse who is attempting to enable individuals, aggregates, and communities to improve their health through collaborative community activity. Although each has elements that foster participation for mutual goal setting, none addresses the power differences that exist between the layperson and the health care professional. The Intersystem Model addresses this problem at the individual level (Artinian, 1991):

> If the patient/client and the nurse are not able to agree upon a plan of care, the nurse will need to decide if she will exercise her expert power to institute a plan of care or if she should attempt another round of communication and negotiations. When either the emotional or physical safety of the patient is in question, she must exercise her professional power to protect the patient. For the nursing process to be effective, the priorities of both intrasystems must be communicated, negotiated, and organized to develop the plan of care. (p. 200)

A basic assumption of the Intersystem Model is that, unless safety issues require a nurse to act unilaterally, the values, beliefs, and traditions of the patient or client take precedence in developing a plan of care. Because of this, noncompliance generally is a nonissue in the Intersystem Model because the plan to be followed is one that the patient or client has developed in collaboration with the health professional.

Although there are some situations involving the safety of the community, such as disasters, for which prompt action would be necessary, thus preventing community collaboration, most of the issues regarding community health are more emergent and could be managed through the negotiation process outlined in the Intersystem Model. By using a participative rather than a directive approach, the values of the community members are honored, they learn problem-solving skills to manage the change process, and the change will be firm and lasting.

Nursing Leadership to Promote Community Participation

In an interactional environment in which the health knowledge, values, and behaviors of a population form the basis of health care planning, a new type of nursing leadership is needed. The nurse must be able to take into account the perspective of the population while at the same time espousing effective health care practices. The nurse must be able to handle the challenges of cultural diversity and the complexities of interactional interconnections. There are also situations in which no single leader is in charge and leadership is shared. Although many large projects require shared leadership, these leadership skills have not been taught. Therefore, new theories and models such as the Intersystem Model are needed to guide this new type of practice. When coordinating the activities of many people who are each partly responsible for acting on complex health care issues, the nurse must be able to inspire and motivate them through persuasion and use negotiation as a tool for bringing about change. An important consideration is the greater power of the health care professional based on position and knowledge, and conscious efforts must be made to ensure that shared leadership and power are an actuality.

Although the community health nurse may have a clearer understanding than the community group about the health goals and the available resources for a particular problem, unless the plan for meeting the goal is developed from the perspectives of the group, nothing will be accomplished. Therefore, the nurse must have the ability to network with those in the group and provide vision to them. The effective nursing leader will be able to provide intellectual leadership to the project by assisting members to sort out and classify significant health issues and identify and evaluate alternative approaches to the problem. The effective leader will also be able to share leadership and develop leadership potential in others.

Providing leadership in community groups can often be a real challenge because the group leader may have no legitimate power. Therefore, leadership rests more on the style and personality of the leader as well as on his or her competence in negotiation and persuasion.

Community building is a skill identified by Conger (1993) that is important to the leader. It involves understanding the issues surrounding diversity and acquiring a tolerance to diverse viewpoints. It involves understanding one's own world and the meaning of the worlds of others as well as the external worlds in which they operate. It also involves being able to think in systems terms so that the pattern of separateness and interconnectivity of individuals, aggregates, and communities is understood. The outcome is achievement of strategic goals by a community dedicated to the project.

In her study of community participation, Yan (1993) identified charac-teristics of a good leader. Respondents said that a good leader was a motivator, very enthusiastic, an energizer, and one who could make others feel success-ful. The leader had the skill of discovering which outcomes the participants valued most and giving them assignments that would realize those outcomes. The leader would also be able to focus ideas.

Community Assessment

Project GENESIS

Because the community health nurse provides nursing care within the broad perspective of the social, economic, educational, cultural, political, and envi-ronmental variables that affect a community, a broad-based assessment plan is necessary to understand the complex relationships of variables influencing the health of a community. A data collection scheme known as Project GENESIS has been developed at the University of Colorado School of Nursing. It is based on the premise that "assessing health needs from the perspective of the members of a community or aggregate is essential to understanding the needs as part of the total picture of life in a community" (Magilvy & Stoner, 1986, p. 1). Project GENESIS focuses on the health needs and quality of life of a community or aggregate. It attempts to assess the usual patterns of leadership, traditions, values, and assets of the community as well as its weaknesses. The goal is to empower citizens to take action in their community.

According to Magilvy and Stoner (1986), "Project GENESIS combines the use of demographic statistics and epidemiology with ethnographic field methods" (p. 1). They explain that "an ethnography is a description and analysis of a total community or aggregate as compiled from the viewpoint of its residents and key leaders" (p. 1). Through participant observation field study, the perceptions of the community members about their health needs and those of the community can be assessed. It is also possible to assess quality of life of the community members, to identify resources such as support groups used by members, and to identify health learning needs of the group. An important part of the project is to then work with the community to help it take action to correct the health problems identified. The five steps involved in carrying out Project GENESIS are to

1. Collect and analyze secondary data (e.g., census, demographics, histories, and public documents)
2. Identify and interview "key informants" (formal and informal leaders in the community)
3. Interview community residents as "primary informants" and conduct participant observations in the community
4. Combine and analyze secondary and primary data—develop lists of strengths, weaknesses, and net health values, resulting in recommendations to improve the health status of the community
5. Present a written and verbal report to the community and ask for feedback (Magilvy & Stoner, 1986, p. 2)

This information will provide a database with which the community nurse can complete an intrasystem analysis of the community as described below and begin to develop joint plans for mutually agreed-on projects.

Intersystem Model Assessment

The Intersystem Model is broad enough in scope to address health problems and their determinants from the perspective of the population. Therefore, it can be used to provide the theoretical framework for collecting and interpreting data and making a community diagnosis. During the assessment process described previously, data are collected about the biological, psychosocial, and spiritual subsystems of the community. Use of the model helps the nurse to focus on the critical aspects of the situation that relate to the situational sense of coherence (SSOC) of the community. By following the steps of data collection outlined in the Project GENESIS guidelines, the community health nurse is able to gain the information necessary to make an assessment of the subsystems of the community in relation to a particular health issue.

Information about each of the subsystems of the community can be collected. At the aggregate level, characteristics of the biological subsystem include observations about such factors as the gender, age, and ethnic composition of the community; level of nutrition and fitness; genetic disabilities; drug dependency patterns in the community; level of health care available; patterns of health care use; and environmental hazards. Characteristics of the psychosocial subsystem of the community include factors such as morale, ethnic traditions, freedom of expression, decision-making patterns, urban or rural mentality, and political orientation. Characteristics of the spiritual sub-system include factors such as church affiliations of community members, degree of religiosity, Eastern or Western philosophical orientation, and religious practices and traditions.

These characteristics affect the interactional processes through which the community organizes its life. A biological characteristic of a community affecting interaction could be the amount of energy available to work on problems. Characteristics of the psychosocial subsystem that influence interaction include factors such as the educational level of the group that could affect the categories available to code information and therefore the ability to process information (the cognitive subsystem); the perceptions, feelings, and emotional patterns of the community that could describe community response to an issue (the affective subsystem); and the community rituals, coping strategies, or traditional practices that indicate the repertoire of behaviors available to a community for response to an issue (the interactional subsystem). Any health care issue is also processed through the spiritual subsystem that is reflected in the worldview(s) of the community and is evidenced by the religious beliefs and values, religious practices, and responses to deity.

A community assessment can also give indications about health outcomes. For example, the level of health knowledge about any health problem can be assessed as well as health awareness in general and knowledge about community policies related to health practices. Community values, such as orientation to health or illness and medical science, and commitment to health practices can be observed. Actual health behaviors can be assessed through studying mortality and morbidity statistics, support of medical care, the percentage of the population with immunizations, use of the health care system, and so on.

This assessment of the community subsystems and intrasystem functioning provides the data for a calculation of the community SSOC about a particular health concern. The community can be scored in terms of health care knowledge, motivation to deal with the health care issue, and community resources on a scale of high, medium, and low. The goal of intervention with the community is to increase community SSOC when it is low. This is done by changing community system characteristics and processing of the event by mutual goal setting. Intervention is evaluated by rescoring the community on community SSOC.

Involving Community:
Case Analysis of a Healthy Cities Project

Hickman (1993) has defined community organization as "the process whereby community change agents empower individuals and aggregates to solve community problems and achieve community goals" (p. 131). It is a joint venture between professionals from many disciplines, including nursing, and volunteers from the community organized in civic groups or social action groups. The goal of community organization is to promote community health. An example of a project to promote health that was undertaken as part of the Healthy Cities initiative will be described. The example is excerpted from the data recorded by Jean Yan for her doctoral research at the University of Southern California.[1] Yan and I worked together to analyze the data to develop the grounded theory of "involving community." What follows is an analysis of the same data within the framework of the Intersystem Model.

Clinical Example

Focus of Situational Interaction

The focal point for interaction between the Healthy Cities Project Steering Committee of Small City, California and community volunteers was the organization of activities necessary to carry out health related projects such as the provision of health information to meet community needs.

Assessment

Developmental Environment

Small City is a suburban city of about 21,000 located in a large metropolitan area. The city has experienced a large growth spurt in the past 20 years, and many of the residents are newcomers who speak a total of 35 different languages. Small City is ethnically diverse with 45% white, 35% Hispanic, 11% Asian, 8% black, and 1% other. Yan found that despite its rapid growth, the city has been able to provide adequately for its residents through fiscal responsibility and imaginative planning necessary to promote the steady upgrading of services. It has received an achievement award from the League of California Cities for "managing resources to carry out public policy in a progressive, timely, and cost effective manner." In 1988, Small City became part of the Healthy Cities Project, and a steering committee was selected to guide the project.

Situational Environment

Intrasystem Assessment of the Steering Committee

Detector Function. A diverse group of individuals made up the steering committee that represented the different interest groups within the city. The steering committee included representatives from the Unified School District, the County Health Department, the chamber of commerce, the Board of Realtors, hospitals, local churches, and a number of voluntary agencies. Each committee member brought a wide range of experience, knowledge, and ability to the group. Therefore, as a group, the committee had knowledge about all aspects of government and health care delivery.

Selector Function. The criteria used for recruiting the steering committee members were that individuals had a known interest in the city and its environment, had visibility in the city from previous activity and involvement in the community, and would be able to commit adequate time to participate actively. Because the members were selected by the project coordinator based on personal relationships and prior experiences of working together on community projects, the values of the committee members may not be representative of the community.

Effector Function. **Organization of the Steering Committee.** Because a major concern in the organization of community health programs is the relative power of community residents and professional health care providers in the planning and implementing of programs, a detailed report of the behaviors of the steering committee in relationship to the community is presented. The goal is to examine the way in which power was handled by the committee in its relationships with the community.

Detailed analysis of the data identified the steering committee to be the focal point for community participation in the project. It was the link between city government and the community. It interpreted the goals of the project to the community and provided a mechanism for residents to discuss health related issues and provide recommendations.

The steering committee was delegated the major task of providing direction for the Healthy Cities Project. It was responsible for drafting goals, developing structures, acquiring resources, and evaluating the effectiveness of the program. The coordinators were responsible for selecting the members, and choices were made based on personal and professional relationships. Relationships within the committee were quite informal, which permitted closer relationships to develop within the group and allowed members to find satisfaction in the group. Yan (1993) wrote, "Their ability to work well together stemmed from their great personal respect and affection for each other" (p. 121).

The committee was structured as an open system and was flexible and cohesive. Members could drop out and join at any time depending on their interest in participating. Meetings were held as often as necessary and the agenda was developed from input from committee members and the particular needs for planning and implementing the various programs. The chairperson of the steering committee was the coordinator of the Healthy Cities Project. This person held the position of Environmental Counsel for Small City. It was the duty of the chair to relay important information and seek support from the city council, strategic planning task force, and volunteers.

Committee Activities: Involving Community. Yan (1993) found that the basic social process of involving community took place in three stages: recruiting, limiting involvement, and sustaining involvement. Yan described these phases as follows:

> The recruiting phase consisted of processes and programs designed to cultivate, generate and extend community involvement in healthy cities' projects. The limiting phase referred to those processes and programs designed to slow down, inhibit, and contain community involvement. The sustaining phase pertained to those processes and programs designed to support, conserve, enhance, and maintain community participation in health programs. The city may be involved in one or two phases simultaneously. These phases were cyclical, interactive, and interdependent. (p. 113)

Recruiting Phase. In the recruiting phase, community residents and committee members were encouraged to share their visions for and opinions about a healthy city. Enthusiasm was high. One resident made the following remark: "This is our community. We are the only people who can do something good for our community. Come and participate. Come, let's do some-

thing worthwhile" (Yan, 1993, p. 129). Strategies such as inform-
ing, networking, marketing, rooting, and providing comfort were
used to recruit community volunteers for the project. Strategies
were developed within the committee, but there was considerable
interaction with community residents.

Limiting Phase. Recruiting strategies were so successful that
many individuals and organizations contributed their views about
how the projects should be designed and carried out. The problem
for the committee was to implement the plan in a timely fashion.
During this phase, committee activity was kept secret from the
community, and the goal was to consolidate power. Yan (1993)
wrote,

> The main challenge was to determine a balance between the extent of
> community participation and effective operation of the Healthy Cities
> Project. To attain this balance, there was a move on the part of the
> Healthy Cities Staff toward "softening participation." "Softening of
> participation" involved processes and programs designed to slow down,
> inhibit, and contain community involvement in the Healthy Cities
> Project. (p. 145)

The softening process focused on limiting the numbers of
participants by weeding out the "talkers," by stopping global
recruitment of participants, by having the staff define the amount
of participation allowed in the project, and by formalizing the
project so that the bureaucratic processes became so complicated
that the average volunteer was happy to turn over the project to
the staff. Rather than the idealism of the recruiting phase, this
phase was marked by control and direction from the staff. An
advisory council was formed and only selected representatives
were allowed to participate. Participation was limited to advice
and recommendations to determine what programs would be of
value to the community. The council had no decision-making
authority. A staff member from the Healthy Cities Project said
that "One of the things we did to soften community participation
was [to create] an advisory body. This limited exactly how far
communities participated in the Project" (Yan, 1993, p. 148).

Participation was also limited by the drafting of bylaws that
described the structure of the project and the composition and
functions of participants. These bylaws provided a set of rules that
ensured that the committee retained control over the project. One

respondent said, "Because of our inability to achieve consensus, we set up bylaws. These bylaws then dictated how far community participation could go" (Yan, 1993, p. 150).

Sustaining Phase. For the project to continue, there needed to be long-term interest from the community. This occurred because community residents saw desired and valued results from the project despite the fact that control of the project had essentially been taken away from them. One community resident said, "People will continue to participate because they believe in the project. They see and experience a better community through the successful implementation of the project" (Yan, 1993, p. 151). The strategies used for sustaining participation were relevancy, recognition, leadership, program setup, rewarding experience, and program ownership. Even during this phase, the committee acted in a demeaning way toward community residents. One respondent explained, "Appreciation is cheap. It costs nothing, but it makes people feel good and encourages them to continue to participate in the project. If you give the participants a little perk, they are back again volunteering" (Yan, 1993, p. 155). Another said, "Recognition is number one for us. They like their names in newsletters, newspapers, TV, etc. It makes them feel good" (Yan, 1993, p. 155).

A major way to sustain participation was to provide good leadership. A committee member said, "Really we have a great group. We have lots of ideas floating around and the coordinator puts them into action. He should be credited for most of the work" (Yan, 1993, p. 157).

Intrasystem Assessment of the Community

Detector Function

Community residents were noted to have little knowledge about the government and programs of Small City—about health services, the senior citizen center, and City Hall. They were also low on health information such as the need for immunizations and where immunizations were available.

Selector Function

Volunteers from the community selected the projects they wanted to participate in based on their own interests, and they felt good about the positive results of the program. Hispanics tended to have little experience in community participation but agreed to participate because of the high value they placed on health. For Asians, family centeredness was the important value, and for them family came before community. They would participate if persuaded that the program would be of help to their families.

Effector Function

During the strategic planning process, health providers and community leaders reported that health and wellness resources were often underused. It was found that city residents were turning to their ministers and social workers for advice on health care. Among the health needs of the community were dental health, smoking cessation, and immunizations for school children. Volunteers for the projects came from all the ethnic groups of the community.

Analysis

Determination of Community Situational Sense of Coherence

Comprehensibility = 1. Health knowledge was determined to be inadequate.

Meaningfulness = 3. Community residents valued health and were willing to participate in programs designed to increase health. It was a community goal to become a healthy city. Through community meetings, residents realized that their health problems were shared by others and believed that the programs would benefit them.

Manageability = 1. Many residents were at a low income level, and the obtaining of health care was difficult.

Community Diagnosis

There was a knowledge deficit about community health resources.

Committee Goals

1. To develop program priorities
2. To implement program goals
3. To decide on future directions of the project

Intersystem Interaction

These intrasystem characteristics formed the milieu in which interaction took place. During interaction, intrasystem information was continually gathered by both community members and steering committee members about what each knew, what values were held, and pertinent behaviors. During intersystem interaction, information was communicated, values were negotiated, and behaviors were organized.

Communication of Information

Communication took place at two levels: information about project activities and actual health information. Initially, much time was spent sensitizing residents to the health needs of the city and what the project might mean to their community. Direct contact, either by door knocking or by phone, was used to invite people to participate in the programs. Successful community meetings were also held, with over 200 community members in attendance. These meetings allowed two-way interaction between staff and residents and many questions could be addressed. Another method was to post flyers in local supermarkets.

Churches also participated in communicating information about the program. In an area with a high proportion of Hispanic residents, the head priest of the Catholic church announced the project in all 11 Masses on Sunday and put an announcement in the bulletin in both Spanish and English.

Strategies of information dissemination were used by the steering committee during the recruiting phase to enhance community awareness and understanding of the project. The committee scheduled meetings in accessible locations at convenient times and was careful to use words the community residents could understand.

Health information was provided to the residents through distribution of the *Wellness Guide* published by the School of Public Health at the University of California, Berkeley, and the Health and Wellness Calendar, which was developed to complement the Wellness Guide. The calendar included a directory of health resources, with information about services available, the emergency and hot line numbers for them, and a list of local community groups.

English and Spanish versions of the guide were distributed free of charge to 8,000 households by bilingual volunteers. The calendar was distributed to all households. Other health-related projects included a Children's Health Fair, a health theme for the annual city picnic, a public opinion survey on attitudes about smoking, an AIDS education forum, and an AIDS play for teenagers.

Negotiation of Values

Health was a major priority for the city at both the resident and official levels, and the city had sponsored a number of health-related projects. The mayor and the city council members wanted health to be a way of life for the people. Therefore, provision of health information and referrals became one of the goals of the Strategic Plan and the Healthy Cities Project. Although health was a shared value, there was little evidence of any negotiation regarding the priorities of the two groups regarding health projects and how to carry them out.

Organization of Behaviors: Mutual Plan

Following a large recruitment drive, more than 100 volunteers participated in distributing the *Wellness Guide* and assisting with the Children's Health Fair, specific programs that had been developed by the committee.

Strategies to sustain participation were to have multiple short-term projects, to set goals that could be accomplished, to provide rewarding experiences that were fun, to let volunteers see the results of the program for the beneficiaries so that the effect on the community could be seen, and to encourage a sense of ownership in the program. Although the mandate of the Healthy Cities Initiative was community participation, it was difficult to find evidence of a plan of action developed jointly between the community and the steering committee. Nevertheless, despite the attempts to limit involvement, some participants experienced a new type of partnership between official health agencies and the community that they had not seen before.

Evaluation of the Plan: Rescoring on SSOC

Comprehensibility = 2. Knowledge levels about available services increased. For example, by using the *Wellness Guide,* a battered wife found shelter for herself and her children, and an immigrant found free prenatal care.

Meaningfulness = 3. The motivation to provide health care for community residents remained strong throughout the program. A steering committee member summarized the attitude of the community: "We are a City of Health and we can't have a healthy city without healthy kids."

Residents said, "I feel encouraged that everybody in the community is serious in making our community a healthy city. Sure I will participate, this is great!" (Yan, 1993, p. 134); "We participate because we know we are helping out friends and our friends' children" (p. 167); and "We have been helped so much in . . . by different things. And we feel that we're giving it back to the community" (p. 167).

> Manageability = 2. Information and health care became more available, but the activities represented episodic events, and continuing programs were needed. One resident with a family income of less that $20,000 per year and no health insurance took his four children to the fair and said that he had saved $289 on immunizations. A Small City councilwoman said that the fair cost between $2,000 and $2,500 but it saved 20 times that much for the residents of the community.

Staff members observed "It's not yet possible to measure how much participation really impacts the community. We can't say it saved x number of lives or that it reduced the number of substance abuses" and "We're not really at the evaluation stage yet . . . but you know that it is making a difference in their lives (Yan, 1993, p. 174).

Discussion

The literature about the California Healthy Cities Project described it as a collaborative, participatory, community-based, coalition-building approach to health development. Although the rhetoric of the project was community ownership of the program, there was little evidence of mutual goal setting and planning in the interactions between steering committee and community. The promise of community ownership was used to get community residents to identify with the project's objectives and goals. Health was seen as an important issue and worthy of active support, but participation at the planning and implementing phases did not take place at the grassroots level—it involved mainly professional residents. One respondent described the new coalition for health (Yan, 1993):

> Our committee is very interesting. Healthy Cities Projects draw not only health professions, but other segments of the community. I'm from education, there are some from business, etc. The health professionals and their determination and drive

pulls me along with them. Their commitment increases my commitment to the project. (p. 164)

Members of the community were seen as people who could augment the services of the public health agencies in times of budgetary problems—"to fill in when the appointed staff no longer have the time to devote to a project." Some residents identified this helping out as "Our way of helping the economy of our city." Although this is noble, participation at the planning stages may have provided more community satisfaction and programs more geared to the needs of the community. The involvement of official agencies contributed to the success of the project, but at the expense of community participation. A respondent noted (Yan, 1993),

> The community definitely could not have done it without the support of City Hall. I guess that I found that City Hall was doing more than what I would have expected. Which is not bad, but I didn't realize that that's their team approach and that's where we begin. So it did surprise me. But I don't think the program could have gone so well had it not had that nucleus. . . . And I don't think that they did the program alone nor did they try to do it alone. (p. 172)

Yan (1993) commented that the limited participation of community in the project might be a reflection of the dilemma of professionals reconciling their disciplinary specialization with the challenges of a new reality requiring more "flexible structure, horizontal relationships, and power sharing" (p. 190). A project such as the Healthy Cities project requires a model for shared leadership with other professionals as well as with the beneficiaries of the program. Yan (1993) wrote, "If participation is to work in the California Healthy Cities Program, it needs to bring together the public, private, voluntary, and community sectors united under the common purpose of promoting health for city residents" (p. 185). The Intersystem Model provides such a framework. The theory of involving community that emerged from this study highlights the problem of the power differences between official and community groups. If the committee is not representative of and responsive to the different interest groups, community participation is in question. Only through an awareness of the problem and a concerted effort to honor the values and input from community groups will it be possible for the community health nurse to engage in mutual partnership with the community he or she serves.

Note

1. Adapted with permission from Jean Yan.

References

Artinian, B. M. (1991). The development of the Intersystem Model. *Journal of Advanced Nursing, 16,* 164-205.

Conger, J. (1993, Winter). The brave new world of leadership training. *Organizational Dynamics,* 46-57.

Cottrell, L. S. (1976). The competent community. In B. H. Kaplan, R. N. Wilson, & A. H. Leighton (Eds.), *Further explorations in social psychiatry* (pp. 195-209). New York: Basic Books.

De Kadt, E. (1982). Community participation for health: The case of Latin America. *World Development, 10,* 573-584.

Goeppinger, J., Lassiter, P. G., & Wilcox, B. (1982). Community health is community competence. *Nursing Outlook, 30,* 464-467.

Hancock, T., & Duhl, L. (1988). *Promoting health in the urban context* (World Health Organization Healthy Cities Papers No. 1). Capenhezen: FADL.

Hickman, P. (1993). Community organization. In J. Swanson & M. Albrecht (Eds.), *Community health nursing: Promoting the health of aggregates* (pp. 130-142). Philadelphia: W. B. Saunders.

Magilvy, K., & Stoner, M. (1986). *Project GENESIS: A description.* Unpublished manuscript, University of Colorado.

Merriam-Webster's Collegiate Dictionary (10th ed.). (1994). Springfield, MA: Merriam-Webster.

Ostrom, V., & Tiebout, C. (1961). The organizations of government in metro areas: A theoretical inquiry. *American Political Science Review, 55,* 831-834.

Swanson, J., & Albrecht, M. (1993). *Community health nursing: Promoting the health of aggregates.* Philadelphia: W. B. Saunders.

Tsouros, A., & Bracht, N. (1990). Principles and strategies of effective participation. *Health Promotion, International, 5,* 199-208.

Yan, J. (1993). *Community participation and the California healthy cities: A grounded theory study.* Unpublished doctoral dissertation, University of Southern California.

11

Curricular Development
Using the Intersystem Model

Phyllis Esslinger
Margaret M. Conger

Concept of Systems
Concept of Development
Concept of Stress
Concept of Situational Sense of Coherence
Course Development
Conclusion

The Intersystem Model was developed to guide the nursing curriculum at Azusa Pacific University (APU). It has provided the structure for curricular development since the inception of the program. The model was initially known as the Nursing Process Systems Model (Brown, 1981). The concepts of individual and family development across the life span, the effect of stressors from both the external and internal environments of the person, and the view of person as composed of subsystems—the biological, psychosocial, and spiritual—were identified as core to the curriculum. In this curriculum, the role of the nurse in the assessment process was to identify stress-related alterations in the client and the nurse to identify personal resources or values that would influence the helping relationship. Emphasis was placed on the interactive process between the client and nurse in developing a mutually agreed-on plan of care that focused on reducing the effects of the stressor.

As the model continued to be developed, the importance of the interactional process between the nurse and client was further emphasized. The model was renamed the Intersystem Model (Artinian, 1991) to reflect this interactive process. Another addition to the model has been the expansion of the definition of the client to include the family, an institution, or a community as part of the concept of person. An important component of the model has been an understanding of both the client's and nurse's subsystems and suprasystems. Both the suprasystem and subsystem will change depending on who the client is. For example, when assessing a family, the subsystems will be individual family members, whereas the suprasystem will be the community.

Also in the model revision of 1991, the concept of health was expanded to include situational sense of coherence (SSOC) as the measurement of a healthy response. With this added conceptualization of health, the role of the nurse was identified as that of assisting the client to make adaptations that would lead to an increased SSOC. The concepts identified in the earlier versions of the model continued to be stressed.

The Intersystem Model has proven effective in providing structure to both curricular and course development in the Nursing School at APU. The students are introduced to the model during their freshman nursing course. This introduction provides the framework for all future analysis of client health problems. The strands of systems theory, developmental theory, and the role of stress in altering health are emphasized throughout the curriculum. Both nursing courses and nursing cognates are developed with these concepts in mind.

Concept of Systems

In the Intersystem Model, the system is the unit of analysis and is a way of viewing a person, family, institution, or community as a whole dynamic entity in a hierarchal relation of subsystems and suprasystems in mutual interaction. The subsystems are all aspects of the system under analysis, whereas the suprasystems are defined as those systems external to the unit of analysis. Thus, if the system of analysis is the person, the family would be its suprasystem. If the family is the unit of analysis, the community would be its suprasystem.

General systems theory as used in the Intersystem Model is introduced in the beginning courses in nursing. It is then integrated in all courses throughout the curriculum. Finally, a specific course in systems theory is provided in the senior year. In this course, the thrust is to provide the student with an understanding of the intricacies of social institutions with a particular focus on those institutions that relate to health and illness. The nurse's role in these social institutions is examined; theories of leadership and management that have an impact on nursing are discussed.

The initial introduction of systems theory focuses on the individual as a system composed of biological, psychosocial, and spiritual subsystems. This individual focus is maintained in the freshman and sophomore nursing courses. Subsequently, in the junior and senior years of the program, the system is also viewed as the family, institution, or community.

The biological subsystem is made up of those components that are physical parts of the body that are grouped together to perform particular functions. Subsystems within the biological system are identified as regulatory (neurological, endocrine, and immune), circulatory and respiratory, musculoskeletal, gastrointestinal, renal, and reproductive. Courses in the basic sciences, including chemistry, anatomy, physiology, microbiology, and pathophysiology, provide the theoretical knowledge to understand this subsystem. Chemistry, anatomy, and physiology are taken during the student's freshman year. Microbiology and pathophysiology are taken concurrently with nursing courses.

The psychosocial subsystem is that which enables the person to relate to self and others. Through interactions with others, one forms perceptions of self and the world. The components of the psychosocial subsystem include affective, cognitive, and interactional functions. The general education courses that provide the theoretical knowledge to comprehend this system include general psychology, human growth and development, cultural anthropology, and sociology. Students receive personal experience in the area of psychosocial nursing through focused interaction with patients and peers and through the university requirement that each student provide community service.

The spiritual subsystem that is central to the person is "the life principle which pervades a person's entire being and which integrates and transcends biopsychosocial nature" (Gordon, 1985, p. 262). This subsystem is based on the assumption that the person's spirit is at the center of being and has ultimate effects on all aspects of life. The components of the spiritual subsystem include the person's relationship to deity, spiritual beliefs, religious practices, and affective responses.

The student is exposed to an understanding of the person's spirituality through 18 general education hours in religion spread over the 4 years of the undergraduate experience. During this time, the student has an opportunity to examine the Christian Western religion as well as other religious traditions. In addition, all students participate in a religious service offered on campus several times a week. All students are also expected to participate in religious community service of their choosing. All these experiences enhance the student's spiritual experience.

Concept of Development

The concept of development assumes that any system or subsystem can have demonstrable differences at various times. These changes can result from directional change in response to environmental stressors, both internal and external. The developmental concept further assumes that "man exists within a framework of development or change"—a life continuum of birth, growth, maturation, death, and beyond. The concept also suggests that "at any point in time, man's present can be seen in terms of his past and potential future" (Brown, 1981, p. 37).

The focus on development as a primary concept in the Nursing Process Systems Model and in the Intersystem Model has had a profound effect on guiding curriculum at APU. After the students have been introduced to elementary nursing skills, the nursing program initiates the study of the beginning family and infant. It then progresses to the study of children and, subsequently, adults. An early course in the curriculum is an introduction to individual human growth and development theory at the sophomore level. This is followed by family development theory at the junior level and community development at the senior level. These courses provide background in developmental theory that prepares students for clinical nursing courses. Figure 11.1 depicts the sequential placement of these courses.

The clinical application of the developmental theory follows the same pattern. Students begin caring for newborns in the sophomore year. The care of individual adult clients also occurs at this level. As the student progresses, emphasis is placed on care of the family. This is followed by a focus on community and the impact of social institutions on health in the senior year. Hence, the curriculum flows from the singular individual with limited health needs to multiple human groupings requiring a complex focus. This progression is also seen in skill development by building progressive levels of difficulty as the student moves though the program.

Concept of Stress

The concept of stress and its effect on person is also an important thread guiding curriculum. Stress is defined in the Intersystem Model as the dynamic interaction (behaviors or responses) between the system (person, family, institution, or community) and the person's subsystems (biological, psychosocial, or spiritual) and suprasystem (environment). This stress is a state of tension in the person that can have a pathological, neutral, or even beneficial effect. Stressors are defined as events, factors, or influences that stem from the internal or external environment as the person (defined as individual or aggregate) passes through time and that provoke a change in behavior.

FRESHMAN

Fall		*Spring*	
55-100 Intro to Nursing Science	1	55-105 Nursing Process: Fundamentals	
52-111 Chemistry for Health Prof.	5	of Professional Nursing	5
51-250 Human Anatomy	4	51-251 Physiology	4
31-110 Freshman Writing Seminar	3	63-110 General Psychology	3
41-112 or 121 Religion (Intro to Bible)	1	64-120 Intro to Sociology	3
74-101 Beginnings: Personal Development	1	— P.E. Elective	1
	15		16

SUMMER

34-111 Public Communication   3 units

SOPHOMORE

Fall		*Spring*	
55-210 Maternal/Infant/Newborn Nsg.		55-210 Maternal/Infant/Newborn Nsg	
-or-	6	-or-	6
55-212 Medical-Surgical Nursing of		55-212 Medical-Surgical Nursing of	
the Adult Client		the Adult Client	
55-220 Health Assessment	3	51-220 General Microbiology	4
55-260 Nutrition	2	63-404 Abnormal Psychology	3
63-290 Human Growth & Development	3	— Religion (Bible)	3
41-208 Religion (Ministries)	3		
	17		16

JUNIOR

Fall		*Spring*	
55-300 Stress Theory in Nursing	3	55-305 Family Theory	3
55-310 Mental Health Nursing (3) *and*		55-310 Mental Health Nursing (3) *and*	
55-313 Restorative Nursing (3)		55-313 Restorative Nursing (3)	
-or-	6	-or-	6
55-314 Medical-Surgical Nursing		55-314 Medical-Surgical Nursing	
of Children (6)		of Children (6)	
63-402 Statistics	3	55-325 Nursing Research	2
— Religion (Bible)	3	55-367 Human Pathophysiology	3
— P.E. Elective	1	64-358 Human Diversity	3
	16		17

SENIOR

Fall		*Spring*	
55-401 Systems Theory in Nursing	2	55-402 Issues in Nursing	2
55-410 Community Health Nursing		55-410 Community Health Nursing	
-or-	6	-or-	6
55-411 Advanced Medical-Surgical		55-411 Advanced Medical-Surgical	
Nursing/Leadership		Nursing/Leadership	
31-111 Introduction to Literature	3	— Religion (Doctrine)	3
— Religion (Ethics)	3	— Fine Arts Elective	3
— U.S. History or Poli. Sci.	3	— Elective	3
	17		17

Figure 11.1. Schedule for Nursing Students Entering Fall Semester 1994

In the Intersystem Model, the environment is defined as all events, factors, and influences that affect persons as they move through time. Whatever is not included in the system (unit of analysis or focus) is the environment. Therefore, stressors can come from the subsystems and suprasystems that are the environment of the system. Adequate resources and coping responses lead to modification of the environment in such a way that the person's SSOC is enhanced. Thus, the person (individual, family, institution, or community) is moved toward optimum health. Inadequate stress response decreases a person's SSOC, thus reducing health.

Although stressors are addressed in all clinical courses, a course specifically designed to address stress theory is offered during the junior year. This course not only provides a theoretical basis for stress but also identifies specific interventions to reduce or change negative responses or to enhance positive responses. Theory about change as a potential stressor is included in both developmental theory courses and stress theory. It is also addressed in depth in the senior year in the course on systems theory.

Concept of Situational Sense of Coherence

The concept of health is defined in the Intersystem Model as a dynamic state of functioning achieved by successful adaptation to stressors through strengthening the situational sense of coherence. This assumption is based on the worldview that a person is a coherent being. Thus, through comprehension, ability to manage, and a sense of meaningfulness, a person can positively adapt to stressors (Antonovsky, 1987).

The goal of the nurse is to assist the client to reach optimum health by making adaptations that reduce the effect of stressors. The result of the nursing action is to assist the client to increase SSOC. This increased SSOC, fostered through the nursing interaction, leads to an improved state of health.

The use of the concept of SSOC allows nursing students to view the continuum of birth through death as separate from the wellness-illness continuum. A person can have an increase in adaptation to an illness situation even though the illness being experienced is chronic or death is the expected outcome. Because the person is both motivated to accept responsibility for the situation and has the necessary resources and knowledge to manage it, SSOC will be increased leading to a higher level of health. This results from providing the person with a sense of control and confidence, hence allowing one to take positive action.

Students are introduced to SSOC during an early nursing course. They become familiar with assessing clients' SSOC in all clinical nursing courses. A standardized care plan that is based on the Intersystem Model is used throughout the nursing courses. This care plan format leads a student through the data collection phase so that an SSOC score for each client can be determined. The interventions planned jointly with the client are then focused to provide the client the opportunity to gain greater control over health management.

Course Development

The Intersystem Model provides the framework with which to structure courses utilizing the concepts of both suprasystems and subsystems (see Figure 11.2). Each nursing course at APU uses the same format as is seen for this junior-year course, the Medical-Surgical Nursing of Children. The course objectives are developed under the headings of unit of analysis and the subsystems of spiritual, biological, and psychosocial as guided by the model. One additional heading, professional concepts, is included that identifies professional, teaching and learning, ethical and legal, and research concepts that are examined in each course.

The use of a universal framework such as the Intersystem Model to provide structure to all the nursing courses in the curriculum is advantageous. The faculty, utilizing the common framework identified by the model, has carefully developed the specifics of content for each course. The format for each course design is such that the nursing concepts and skills incorporated into each course can be traced throughout the curriculum. This ensures against repetition of content from course to course and allows for building on concepts and even identifying missing content. In addition, it is possible to follow the development of a concept as it is presented in an increasingly more sophisticated manner in courses throughout the curriculum.

Nursing Process: Care Plan

In addition to the use of the model to structure didactic courses, it is also used to guide the student through the nursing process. At APU, the goal of the nursing process is to move the client toward optimum health by assisting the client to reduce the effects of stressors and thus increase SSOC. When the nursing interventions have been completed, an evaluation of the effectiveness of the nursing process is done by reevaluating the client's SSOC. The new SSOC score is compared to the initial score and provides the basis for evaluating the effect of the nursing interventions. Thus, a quantitative comparison between the client's initial SSOC and final SSOC can be made and is indicative of the efficacy of the nursing interventions. The changes in SSOC reflect the movement of the client along the health continuum. Use of the

(text continued on p. 203)

COURSE TITLE: Medical-Surgical Nursing of Children

COURSE CREDIT AND CLOCK HOURS: 6 units, 28 hours, lecture/discussion
168 hours clinical

PLACEMENT IN CURRICULUM: Junior Level

FACULTY: Jan Chandler, RN, MSN

CATALOG DESCRIPTION:
This course has both theory and a clinical component in which students focus on the care of
children from birth through adolescence. Emphasis is placed on the effect of illness on growth
and development during childhood. The students care for acutely ill children in a hospital
setting during which time normal growth and development needs during illness and
hospitalization are emphasized. Experiences and data are presented regarding care of children
with special health problems. Prerequisites: 55-212, 63-290.

COURSE OBJECTIVES:
I. Unit of Analysis: Provides comprehensive nursing care to children within a family context.
II. Subsystems:
   A. Spiritual System
      1. Promotes hope within family.
      2. Identifies needs for forgiveness when appropriate.
      3. Promotes family cohesiveness and support during hospitalization.
      4. Analyzes varying religious expressions of family members.
      5. Assesses the impact of childhood illnesses on the spiritual beliefs and practices of
         children and families, developing appropriate nursing interventions.
      6. Relates the moral developmental level of children to nursing interventions in spiritual
         domain.
   B. Biological System
      1. Processes based on nursing diagnosis focusing on problems primarily unique to
         childhood relative to the following areas:
         a. abuse
         b. developmental disabilities
         c. infectious diseases
         d. circulatory and cardiac deviations
         e. hematologic deviations
         f. respiratory deviations
         g. digestive deviations
         h. integumentary deviations
         i. genital-urinary deviations
         j. musculo-skeletal deviations
      2. Applies physical assessment skills consistently, adapting them to the child.
      3. Compares the physical growth and developmental level of the well and sick child,
         recognizing the individual differences in rates of growth in children of the same age.

Figure 11.2. Sample Course

4. Comprehends importance of routine restorative interventions such as immunization status and health maintenance care.
5. Adapts prior nursing skills to the appropriate developmental level of the child.
6. Promotes safety and accuracy in performing and recording the following skills in nursing care of children:
   a. administration of medications
   b. I.V. maintenance
   c. aseptic techniques
   d. isolation techniques
   e. hygienic care
   f. patient airway techniques (e.g., tracheostomy care and suctioning)
   g. collection of specimens
   h. cooling measures

C. Psychosocial System
   1. Processes and implements, based on nursing diagnosis, problems primarily unique to childhood relative to the following areas:
      a. death and dying
      b. separation and loss
      c. hospitalization/illness
      d. psychosocial development
      e. cognitive development
      f. language development
      g. fears
   2. Assesses the psychosocial developmental levels of children.
   3. Communicates with a child at the appropriate developmental level.
   4. Initiates play therapy to assist the individual child with fears and anxieties of hospitalization.
   5. Establishes trust and rapport with the child.
   6. Relates to parents with respect as advocate for child and the health care system.
   7. Implements appropriate ways to prepare a child for procedures and surgery.
   8. Assists the child and family to adapt to illness.
   9. Includes family members as an integral aspect of child abuse nursing care.
   10. Explains common coping mechanisms used by hospitalized/ill children.
   11. Demonstrates understanding of stress in areas of loss and grief, sexuality, cultural-ethnicity, and socio-economic status relative to the illness of the child and the family structure/function.
   12. Initiates means for reducing psychosocial stress in the ill child and the family.

III. Professional Concepts:
   A. Professionalism
      1. Gathers pertinent information through physical assessment data, observation of the child, interview with family members, discussion with health team members, and chart information.
      2. Establishes appropriate nursing diagnosis based upon a comprehensive assessment, scientific nursing principles, and developmental theory.
      3. Establishes long and short term patient objectives based on nursing diagnosis and developmental level of the child.

Figure 11.2. Sample Course, Continued

4. Implementation:
    a. Carries out a plan of care for a child relative to the developmental level.
    b. Involves the child's family in the implementation of nursing care and provides appropriate health teaching.
    c. Provides for continuity of care through documentation and communication with appropriate health team members.
5. Begins to prioritize nursing diagnoses and associated nursing interventions.
6. Begins to select professional experiences for growth.
7. Functions as a full member of the health care team.
8. Discusses the historical evolution of maternal-child health.

B. Teaching/Learning
    1. Adapts teaching/learning theory to appropriate developmental levels of child and family.

C.,D. Ethical/Legal
    1. Identifies ethical issues in the care of children.
    2. Incorporates knowledge of legal issues into nursing practice.
        a. Implements the ANA standards of pediatric nursing practice.
        b. Discusses legal implications of child rights laws and child abuse laws relative to the legal implications for nursing practice.
        c. Acts as an advocate for children.

E. Research
    1. Discusses current research in child care.

Figure 11.2. Sample Course, Continued

SSOC has proven to be an extremely valuable teaching tool. It not only provides the student with an objective means of assessing the client's movement toward health but also provides a framework for identifying areas of client need for further nursing interventions.

The evaluation of the client's SSOC is based on an assessment of the three dimensions of the SSOC: comprehensibility, meaningfulness, and manageability. Each of the dimensions is evaluated on a scale from 0 to 3, the value 0 being low and the value 3 being high. The intent is to assist the client to move toward the high end, optimizing health in the current situation. An overall value for the SSOC can also be obtained by adding the values for each dimension, with the client receiving the maximum score of 9 or a minimum score of 0. The Intersystem Model is not limited in its scope of application. It can be used equally as well with a client requiring acute hospital care as with those clients seeking health care in community settings. It can also be used with a family, an institution, or a community. The thought processes developed as the student moves through the steps of the model work equally well in all settings. Thus, the model provides a solid framework for students to learn the nursing process in a variety of settings.

A nursing care plan format has been developed to guide the student through use of the model (see Figure 11.3). It is used in all clinical courses throughout the curriculum. The care plan includes all the elements of the model. The students are guided to perform an intrasystem assessment of both the nurse and the client. Subjective and objective behavioral data are collected from both client observations and written records. Both stressors and coping responses present in the situation are identified. Any areas of discordance between the nurse values and client values are also identified. Finally, the client is scored on the SSOC and appropriate nursing diagnoses are developed. In using the nursing care plan format, students are instructed to leave blank any sections for which data are not available.

Collaboratively, the nurse and client develop goals for the client. These include both short- and long-term goals and have estimated times for achievement. The nurse will then develop nursing orders in conjunction with physician orders for the client. The orders are specific to both when and where the interventions are to occur. Because the nursing process is reiterative, as new assessment data are identified, changes in these goals occur.

The care plan format is modified for students in the introductory nursing course. These modifications include identification of a single client stressor, a single nursing diagnosis, and evaluation of the client based on the SSOC score changes. These modifications make the care plan form less sophisticated for student use at this beginning level. This care plan format has been modified to provide a shortened narrative format for use by advanced-level students. An example of this format is provided in Chapter 8 as a "clinical example." This format places more emphasis on a narrative account of the interaction between client and nurse.

Effect of the Intersystem Model on Curricular Development

The Intersystem Model has been an effective tool to assist faculty in structuring curriculum and courses at APU. It has also proven to be effective in leveling student nursing experiences and knowledge throughout the nursing program. Also, the model has provided a workable framework with which to guide students through the nursing process. This instrument has proven to be successful in guiding student identification of patient problems and appropriate nursing interventions. The emphasis on a quantifiable tool to evaluate changes in a client's health resulting from participation in the nursing process is valuable.

The model also emphasizes the importance of viewing each patient or client as an individual. This stress on adapting nursing interventions to the specific needs of a particular client fits in well with the curricular changes advanced by Bevis and Watson (1989). An area of learning that they have suggested is important at the baccalaureate level is that of syntactical learning—the arrangement of data into meaningful wholes. This type of learning promotes

(text continued on p. 206)

Intrasystem Assessment

Intrasystem Assessment of Nurse	*Intrasystem Assessment of Patient*
Nurse behaviors	Patient behaviors (subjective and objective)
Biological subsystem	Biological subsystem
Beliefs/knowledge	Beliefs/knowledge
Values/attitudes	Values/attitudes
Behavior manifested	Behavior manifested
Psychosocial subsystem	Psychosocial subsystem
Beliefs/knowledge	Beliefs/knowledge
Values/attitudes	Values/attitudes
Behavior manifested	Behavior manifested
Spiritual subsystem	Spiritual subsystem
Beliefs/knowledge	Beliefs/knowledge
Values/attitudes	Values/attitudes
Behavior manifested	Behavior manifested

Assessment   **Diagnosis**

Scientific basis for patient behavioral data	*Stressors*	*Scientific basis for stressors*	Nursing diagnosis A. Sense of coherence (adaptation) Comprehensibility (score 0-3) Meaningfulness (score 0-3) Manageability (score 0-3)
	Coping responses	*Scientific basis for coping response*	B. Actual nursing diagnosis
	Tentative nursing diagnosis/diagnoses		

Planning   **Intervention**

Patient goals Short term: Long term:	*Nursing orders*	*Scientific basis for interventions*

Figure 11.3. The Intersystem Model Nursing Process

Evaluation

A. Interventions	B. Sense of coherence Comprehensibility (score 0-3)	D. Modification of nursing goals & interventions
	Meaningfulness (score 0-3)	
	Manageability (score 0-3)	
	C. Overall evaluation of patient goals	

Figure 11.3. The Intersystem Model Nursing Process, Continued

variation from the rules or ordinary expectations so that nursing care is based on the unique qualities of the situation. The Intersystem Model, with its emphasis on the uniqueness of both the client and the nurse in a particular situation, fosters syntactical learning. In the clinical area, the student can be guided to search for and find new meanings in the client-nurse interaction. This results in seeing the clients as whole persons responding to the contextual settings in which they are placed.

The model has also been useful in developing the self-study document required for preparation for a National League for Nursing (NLN) accreditation visit. In this process, each concept of the model was easily traceable throughout the curriculum because the concepts were addressed in all course and program objectives. This allowed for rapid identification and analysis of the educational process and student outcomes.

The model has also been shown to provide a framework on which nursing graduates can develop their professional practice. Graduates of the APU nursing program have frequently commented that they continue to use the Intersystem Model as a framework for their nursing practice. Students in the Master of Science of Nursing program at APU have also commented on the ease with which they have learned to apply the model in their clinical practice. They have especially found the focus on identifying the unique qualities of each client to be a rewarding approach to the practice of nursing. Thus, the model has shown its usefulness as a practice model as well as providing structure for curriculum.

Conclusion

The Nursing Process Systems Model that became the Intersystem Model has guided both curricular development and student nursing practice at APU since the inception of the nursing program. During this time, curricular changes have occurred to keep pace with changes in the health care delivery system. With each change, the model has provided the framework necessary to maintain the integrity of the nursing program. As nursing programs continue to evolve to meet the health care system changes mandated in Nursing 2000, such structure is vital. The movement of health care into community settings with a focus on families and communities as well as individuals, is a new challenge to nursing faculties. Because the Intersystem Model works well with communities or institutions as clients, as well as with individuals or families, it can serve as a structure for this change. Thus, the Intersystem Model lends itself well not only to all aspects of current nursing education but also to future developments in education. It is dynamic and flexible, and it provides a rational, concrete, and organized manner for approaching nursing practice.

References

Antonovsky, A. (1987). *Unraveling the mystery of health: How people manage stress and stay well.* San Francisco: Jossey-Bass.

Artinian, B. (1991). The development of the Intersystem Model. *Journal of Advanced Nursing, 16,* 194-205.

Bevis, E., & Watson, J. (1989). *Toward a caring curriculum: A new pedagogy for nursing.* New York: National League for Nursing.

Brown, S. J. (1981). The nursing process systems model. *Journal of Nursing Education, 20,* 36-40.

Gordon, M. (1985). *Manual of nursing diagnosis: 1984-85.* New York: McGraw-Hill.

12

Learning Theories Used in the Application of the Intersystem Model in Nursing Practice

Margaret M. Conger
Ilene M. Decker

Part I: Learning Theories Used in Nurse Development
Part II: Learning Theories Used in Client Care
Conclusion

The Intersystem Model is based on the developmental concept that a person (system) changes as a result of interaction with the environment. The concept of health as defined in this model also incorporates the idea that successful adaptation to environmental stressors promotes health. A number of theories that describe how a person learns also incorporate the idea of learning as a positive interaction with an environmental stimulus. The purpose of this chapter is to examine the learning theories that are useful in understanding this person-environment interactive process as developed in the Intersystem Model.

The two systems that compose the Intersystem Model are the person or client system (individual, family, community, or institution) and the nurse system. For the client and the nurse to work interactively, both need to acquire new skills, knowledge, and affective responses. In addition, the nurse needs to learn effective data collection and interpretation skills, decision-making skills, and motivational strategies. This chapter will focus on learning theories relevant to information processing in two areas: (a) learning theories that guide the instructional process as nurses develop skill in use of the Intersystem Model and (b) learning theories used by nurses while caring for clients within the framework of the Intersystem Model.

Part I: Learning Theories Used in Nurse Development

Gagne: Cognitive Learning Theory

Before looking at learning theories that apply to the use of the Intersystem Model, a general overview of the components of learning is needed. Gagne (1984, 1985) has developed a theory of learning that is widely accepted. It includes a set of cognitive activities that make use of information processing to transform an environmental stimulus into a new capability. This information processing model incorporates five separate domains of learning: the use of verbal information, intellectual skills, motor skills, cognitive skills, and attitude. Table 12.1 depicts each of these skills and how they are used in learning. Verbal information skills incorporate the encoding of information into words for storage that can later be used for higher-order learning activities. Intellectual skills are closely correlated with verbal information and use symbols, such as numbers, letters, or words, to interact with the environment or to make decisions about the environment or both. Motor skills involve the manipulation of objects or the use of speech.

A higher order of learning involves the management of learning, in which cognitive skills are used (Gagne, 1972). Cognitive strategies control the management of learning such as remembering or thinking. Critical thinking is one component of this domain of learning. Finally, attitude determines the choices a person makes about learning and modulates a person's response to learning.

In addition to these domains of learning, Gagne (1985) stresses the importance of both internal and external conditions of learning. The internal state consists of prerequisite skills and attitudes that will influence the new learning. The external state incorporates stimuli in the environment that support the internal learning processes.

When the nursing student is introduced to the use of the Intersystem Model, the learning environment must be carefully considered. The environmental stimuli that the teacher or nurse creates will become the event of instruction. The type of instruction planned must be appropriate to the learning desired. That is, the same instructional techniques cannot be used for all five domains of learning. With cognitive learning, repeated opportunities for problem solving are needed. When the domain of learning desired is attitude, an instructional approach affecting the learner's feelings is appropriate.

Table 12.1 Categories of Learning—Gagne

Category	Description	Outcome	Example: Patient Requiring Diabetic Teaching
Verbal skills	Retrieves from stored memory information such as words, numbers, facts	Uses words to define, state, or clarify an idea	Patient states the important elements of diabetic management
Intellectual skills	Uses mental operations in response to environmental stimuli	Discriminates among symbols, identifies concepts, utilizes rules to predict outcomes, and solves problems	Patient discriminates between NPH and regular insulin
Motor skills	Performs a sequence of physical movements	Executes a series of actions leading to a desired outcome	Patient withdraws insulin from a vial
Cognitive skills	Manages control processes that are needed to think and learn	Develops methods to attend to stimuli, encode information for memory storage, and shift strategies for problem solving	Patient develops a plan for a daily schedule for insulin administration
Attitude	Influences that determine a person's choice about actions to be taken	Chooses an action to take in response to a stimulus	Patient decides to carry out a daily plan for blood sugar management

When designing a curriculum to teach the use of the Intersystem Model, application of Gagne's theory of learning can be used. The first step is to develop goal statements that include specific behavioral objectives. Much of the emphasis on behavioral objectives in nursing curriculum has its origin in Gagne's work (Bevis & Watson, 1989). Following the development of the behavioral objectives, specific instructional steps are identified that are appropriate to the desired domain of learning. Finally, instructional activities that will enhance the transfer of learning from the specific situation to a more general application are needed.

Bevis and Watson

Not all nursing educators are in accord with the stepwise process that evolves from the use of behavioral objectives. Bevis and Watson (1989)

Table 12.2 Typology of Learning: Bevis and Watson

Name	Description	Level
Item learning	Learns pieces of information	Fundamentals course
Directive learning	Uses rules for decision making	Novice learning
Rationale learning	Provides rationale for nursing actions	Novice learning
Contextual learning	Learns mores, rituals of nursing	Throughout student clinical experience
Syntactical learning	Provides nursing care based on the unique qualities of the situation	Clinical experience
Inquiry learning	Generates nursing theory	Graduate school and advanced practice

suggest that overreliance on them may result in the loss of the essence of nursing. Other paradigms are needed in nursing education. They believe additional domains of learning that focus on the analysis of a given situation and generation of new information are needed in nursing curriculum. They also suggest that strict reliance on specific behavioral objectives leads to rote learning and hinders future thought development.

Bevis and Watson (1989) have developed a typology of learning consisting of six distinct stages (see Table 12.2). These stages are not necessarily sequential. Included in the typology are

1. Item learning—in which pieces of information are learned to assist in performing tasks or procedures such as those found in a fundamentals of nursing course.
2. Directive learning—in which rules, expectations, and the exceptions to the rules are used. Much of novice nurse learning is at this level (Benner, 1984).
3. Rationale learning—in which the reasons for what is done are learned. This stage provides a foundation for what is known about nursing practice.
4. Contextual learning—the stage in which the cultural framework for nursing is learned, including the mores, rituals, and accepted ways of being a nurse.
5. Syntactical learning—in which data are arranged into meaningful wholes. This stage allows the learner to vary from the rules or ordinary expectations to provide nursing care based on the unique qualities of the situation.
6. Inquiry learning—in which the art of investigation or theory generation is learned.

Although all these stages of learning are important at various points of a nurse's development, the syntactical learning stage is necessary in learning the use of the Intersystem Model. The emphasis in this model is on the uniqueness of both the client and the nurse in a particular situation. The nurse must learn to assess both the client and nurse situation, make decisions about the value system of each, and develop interactions appropriate to the unique situation. These activities are best described as syntactical learning in the Bevis and Watson (1989) typology.

Bevis and Watson (1989) suggest that at the baccalaureate level of nursing at least 15% of the planned learning activities should be within this stage. They also suggest that there is virtually no planned curriculum in current baccalaureate programs at this level. Thus, it is suggested that adoption of the Intersystem Model would be a significant step in developing curriculum that would support this type of learning. They also suggest that syntactical learning takes place best in the clinical area. In this arena, the student can be guided to search for and find new meanings in the patient-nurse interaction. This clinical learning provides an opportunity to crystallize insights and to evaluate, project, predict, and intuit meanings to the situation.

Critical Thinking Theory

The need for critical thinking in cognitive learning is emphasized in the work of both Gagne (1980) and Bevis and Watson (1989). The constructs incorporated in critical thinking that are especially important to teaching the use of the Intersystem Model to nursing students are reviewed below.

A recent definition of critical thinking (Facione, 1990) reflects the work of a number of educators. This definition incorporates the following cognitive skills: interpretation, analysis, evaluation, inference, explanation, and self-regulation. The use of these constructs is necessary to the nurse to address problems in decision making about client care, consider alternative actions, and make decisions about what to do. Facione and Facione (1992) have also described the elements that increase a person's disposition to think critically. These include

1. Inquisitiveness—a measure of intellectual curiosity and desire for learning
2. Systematicity—use of an orderly, focused, and diligent process in the inquiry stage
3. Analyticity—use of reason and evidence to resolve problems
4. Truth seeking—honesty and objectivity with findings even if they do not support one's own beliefs
5. Open-mindedness—tolerance of divergent views
6. Critical thinking self-confidence—trust in one's own reasoning powers
7. Cognitive maturity—recognition that some problems have more than one option

Older definitions of critical thinking that are also relevant to the application of the Intersystem Model are presented in Table 12.3. Watson and Glaser (1964) interpret critical thinking as an attitude of inquiry that includes recognition of the existence of a problem, the knowledge to analyze the accuracy

Table 12.3 Critical Thinking, the Nursing Process, and the Intersystem Model

	Reference			
Watson and Glaser (1964)	*Ennis (1985)*	*Kurfiss (1988)*	*Nursing Process*	*Intersystem Model*
Define a problem Select pertinent information	Focus on a question Analyze arguments Judge credibility of a source of information	Identify an issue Consider the data relevant to the issue	Assessment	Detector phase
Recognize stated and unstated assumptions Formulate hypotheses	Deduce and judge deductions Induce and judge inductions Investigate value judgments Infer hypotheses	Generate plausible hypotheses	Develop nursing diagnoses	
	Decide on an action	Develop procedures to test the hypotheses	Plan of action	Selector phase
	Interact with others	Carry out the hypothesis testing	Carry out the interventions	Effector phase
Draw conclusions Judge the validity of inferences		Articulate the results of testing Revise initial hypotheses	Evaluate the response to interventions	

and logic of the supporting data, and the skill in applying this attitude and knowledge.

Ennis (1985, p. 45) describes critical thinking as "reflective and reasonable thinking that is focused on deciding what to do or to believe." The correct assessment of statements and formulation of hypotheses, questions, alternatives, and plans of action are necessary skills for appropriate decision making.

Kurfiss (1988, p. 2) depicts or defines critical thinking as an "investigation whose purpose is to explore a situation, phenomenon, question, or problem to arrive at a hypothesis or conclusion about it that integrates all available information and that can therefore be convincingly justified."

Theoretical knowledge and experience in a discipline are needed to apply critical thinking skills. Application of the Intersystem Model provides a context in which critical thinking skills can be used in making patient care decisions based on theoretical knowledge. The clinical area provides the environment in which this can be accomplished most readily. The instructor role is to assist students to reach the syntactical level of learning. A suggested approach to achieve this in the clinical area is depicted in the model shown in Figure 12.1, in which student activities are shown on the left and instructor activities on the right. With the interchange between students and instructors shown in the model, the use of critical thinking is enhanced.

During the detector phase, the student begins to synthesize information from client assessments to identify the client's level of situational sense of coherence and to formulate nursing diagnoses. Through dialogue with the instructor, the student identifies additional information needed to refine hypotheses as necessary. This process enables the student to achieve a higher level of data analysis and arrive at a more explicit hypothesis.

Clinical Application of the Critical Thinking Model in the Intersystem Model

The case that follows incorporates the use of the algorithm during the detector phase and has been derived from an interaction between a student and one of the authors during a clinical instruction experience.

A senior nursing student was caring for a 65-year-old female, Betty Smith (pseudonym). She was admitted with several broken bones in her foot and ankle sustained while hiking alone in sandals over rough desert terrain. Following a surgical procedure to stabilize bones, her physician informed her that she would need to go to a nursing home while she regained her mobility. At the beginning of the interaction between the nursing student and the patient, the student saw no other alternative for follow-up care for Ms. Smith than the nursing home as suggested. The initial nursing diagnosis developed by the student was impaired mobility related to the several fractures in her foot and ankle. The intervention proposed was a period of time of rehabilitation in a nursing home.

The instructor reviewed the data with the student, integrating concepts from the Intersystem Model. The student was encouraged to explore the patient's value system. Through a more in-depth assessment, the student identified independence to be extremely important to Ms. Smith. Another potential problem was depression related to the fear of becoming an invalid and not being able to return to an active lifestyle that included long rambles through the desert. A third area of need identified was the need for education about

STUDENT   INSTRUCTOR

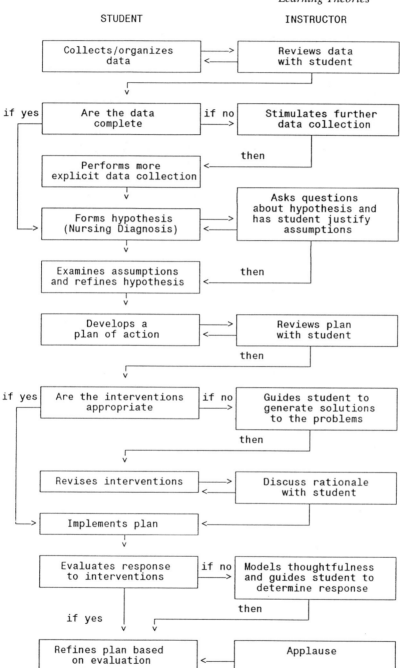

Figure 12.1. Algorithm: Fostering Critical Thinking

appropriate footwear when outdoors. The student was able to justify each of these assessments with appropriate data.

Using these data, the student was now able to develop other nursing diagnoses. In addition to the immobility problem already identified, altered health maintenance related to an unacceptable discharge plan and insufficient knowledge about proper precautions for desert hiking were recognized.

These identified problems were examined more closely through student-instructor dialogue. The original plan of action for discharge to a nursing home was reevaluated. The instructor encouraged the student to further assess the postdischarge possibilities for Ms. Smith. Was the time in the nursing home the only alternative for her? What family or neighbor support systems were present? What possibilities of home health assistance were present? The student worked with Ms. Smith to develop other postdischarge options. As the student and patient worked together, Ms. Smith's demeanor changed radically. She moved from being depressed and crying to becoming very animated in exploring alternate possibilities for her postdischarge care. Together, she and the student were able to develop a plan for her to return home with the assistance of a niece who was happy to come and spend 2 weeks with her. Physical therapy and home nursing visits were arranged to manage her medical needs. In addition, once Ms. Smith found someone who was willing to listen to her and work with her to explore alternatives to the nursing home plan, she began to cooperate more with her physical therapy regimen and quickly learned to use crutches. She changed from an angry, uncooperative patient to one who actively began to take charge of her health. She became very excited over the prospect of once again resuming her active lifestyle. She was even willing to talk with the student about appropriate footwear for hiking in the terrain near her home and made a commitment to purchase hiking boots as soon as she was able to walk again.

Conclusion

In this example, the need to apply critical thinking skills during the detector phase is apparent. By applying the steps of the model, the instructor is able to guide a student in critically analyzing a client situation. The cognitive skills of data interpretation and evaluation are needed to move beyond the superficial nursing diagnosis of simple immobility. The meaning of this immobility to Ms. Smith was the more critical issue. It is through student-instructor interaction that the student learns to evaluate the specific data collected and find meaning in the entire situation. The level of expertise that will be achieved through this process will vary among students.

Part II: Learning Theories Used in Client Care

Detector Phase

Although Gagne's work forms a foundation for understanding how the human learns in a general sense, a number of other learning theorists are important when considering the type of learning to carry out the various phases of the Intersystem Model. In the detector phase of the interaction between the nurse and client, the focus is on making assessments about the client's ability to comprehend, manage, and find meaning in the situation. Using the constructs of the situational sense of coherence, the nurse evaluates the client's knowledge about the problem, his or her ability to manage the problem, and his or her motivation to participate in its management. The client's learning style must also be assessed. The cognitive learning theories previously discussed are needed to carry out this phase.

Selector Phase

The learning theories relevant in the selector phase are most applicable to the nurse-client interaction. In the selector phase of the interaction, the nurse and client work together to clarify the value systems important to each. The nurse's knowledge of several learning theories is important when identifying a client's value system. Where there is discordance between the value systems of the nurse and client, the nurse must be careful to view the situation from the client's point of view. Nursing diagnoses developed and a subsequent plan of care must incorporate the client's value system. Learning theories that provide insights needed to assist clients in clarifying value systems include Bandura's (1986) social-cognitive theory and Weiner's (1985) attribution theory. At times, clients may need help to alter behaviors so that their value systems can be lived out. At other times, clients may need assistance to develop new values based on changes in health status.

Bandura Social-Cognitive Theory

Bandura (1978) has emphasized the influence of environment on behavior. He believes that there are three factors that influence a person's responses. He sees this as a triangle consisting of the behavior, personal factors of the observer, and environmental influences (Figure 12.2). The personal factors that influence the learning environment include characteristics such as race, sex, and appearance. In turn, the way a person is treated in the learning environment will affect perception of self.

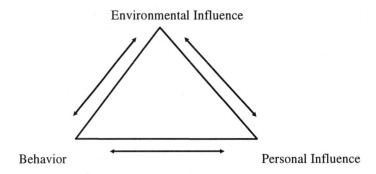

Figure 12.2. Bandura: Influences on Personal Responses
SOURCE: Adapted by Conger from material in Bandura (1978).

The relationship between the environment and behavior is also bidirectional. The way people perceive the learning environment will have an effect on their behavior, and their behavior can alter the environment. There is also a bidirectional relationship between personal factors and behavior. One's expectations and values influence one's behavior, and one's behavior will generate a new concept of self. This complex interaction among person, environment, and behavior must be considered when planning an educational experience.

Bandura's social-cognitive theory also differs from a behaviorist view of learning in reference to what new learned behaviors a person exhibits. Bandura (1971) suggests that a learner decides what observed behaviors to demonstrate. The learner can acquire verbal and visual codes of behavior that he or she may choose not to demonstrate.

This ability of the learner to choose what behaviors to demonstrate can make the evaluation of learning somewhat difficult. A commonly held belief is that learning can be evaluated by a person's demonstrated behavior—either verbally or with psychomotor actions. In light of Bandura's (1965, 1971) findings, however, one must consider that a person may have learned a new behavior but is not ready to demonstrate it; thus, it may be difficult to evaluate the client's learning at that time.

Another aspect of learning described by Bandura (1986) is the effect of role models on determining thought patterns, values, attitudes, and styles of behavior of a person. He suggests that a learner can abstract a range of information from the behavior of others and make a decision about which of the observed behaviors to enact. An application of this modeling behavior is the use of role models to assist a client to alter his or her value system. A positive role model can have a tremendous influence on the client's learning.

When working with a client who has had a surgical procedure such as a colostomy, linking the client with another person who has learned to manage this health problem may be helpful. The client may develop coping strategies to make a successful adaptation to the colostomy through identification with the role model. Support groups, such as "Reach for Life" for women following mastectomy, laryngectomy groups, and others, can all provide important role models for the newly diagnosed client.

Weiner Attribution Theory

Attribution theory as developed by Weiner (1985) explores the influence of both personal and environmental controls on the learning situation. Weiner believes that a primary motivational force for a person's learning is a search for understanding why events occur. He says that each person attempts a cognitive analysis of positive and negative outcomes of previous life events and that future behavior will be influenced by the person's belief system about the causation of these events.

Personal beliefs about the cause for success influence the degree of motivation for learning (Weiner, 1985). Learners who believe that they have personal control over their destiny are more likely to take an active role in learning. Those who believe that luck, fate, or powerful others are the reason for success are less likely to take responsibility for making changes needed to improve a situation. Thus, whatever the nurse can do to help the client retain control over health maintenance will enhance the learning. As the nurse supports the client's locus of control, there is an increased likelihood that the patient will be motivated to actively participate in learning.

A person's potential for learning is often judged by another person by the effort the learner puts into the situation. Weiner (1980, 1982) suggests that when a person perceives that another is not putting much effort into managing a situation, the viewer tends to develop a negative reaction to the person. This is important for the nurse to realize when working with a client. The client who appears to put little effort into managing a health problem can elicit a negative reaction from the nurse. When the value systems of the nurse and client are in such conflict, it is easy for the nurse to withdraw support from the client.

Well-defined learning goals will increase the likelihood of success in skill and problem-solving development. If a person bases perceived learning needs on social comparisons—a comparison with others—the level of success will be lower. Thus, the more the learner can develop goals that relate to solving individual needs rather than comparison with others, the more positive the learning outcome will be (Ames & Archer, 1988).

Effector Phase

During the effector phase, the nurse is often faced with the task of assisting the client to change a behavior. Some of the learning theories developed within the behaviorist tradition as well as those explicated by adult learning theorists are important in nurse-client interactions during the effector phase.

Behaviorist Theories

The learning process as perceived by the various behaviorist theorists revolves around the concept that learning is determined by a change in behavior (Gredler, 1992). In operant conditioning (Skinner, 1953, 1989), a discriminative stimulus is given, the learner responds to the stimulus, and a consequence occurs. This pattern follows the work of earlier behaviorists such as Pavlov, who described the stimulus-response (S-R) mode of learning. Skinner differed from Pavlov in that he believed that the learning arose from the consequence of the action taken and not the stimulus itself (Gredler, 1992).

The consequence of an action can act as either a positive or negative reinforcer (Skinner, 1953, 1989). In positive reinforcement, the consequence of the action results in a new stimulus. In negative reinforcement, the consequence of the action is the removal of a stimulus. Both of these reinforcers have important effects on behavior. In positive reinforcement, the consequence of the action results in a stimulus that the learner finds to be pleasant or helpful. In negative reinforcement, the consequence of the action is the withdrawal of some undesirable stimulus. Both positive and negative reinforcement have been used when one wants to alter a behavior in another person.

Many weight-reduction programs use a behavior modification process. For example, a client may not experience weight loss for several weeks during a weight reduction program. At the same time, however, the person may experience a decrease in clothing size. The nurse can reinforce the importance of this positive outcome. The effect of clothing size change can serve as positive reinforcement for further weight-reduction behavior.

Educational Activities

An educational activity is often the appropriate intervention to assist a client in enhancing health. For this to be most successful, the nurse must understand the client's learning style. The clients that nurses serve are distributed across the life span from the youngest premature infants to geriatric clients facing their final days. In each client situation, the nurse is expected to identify the decision maker for the situation. With the pediatric patient, the parents are most often the decision makers. The child will be brought into the decision process based on development level. With the older person, often the adult children become involved in the decision process. At times, the decision making shifts to a community source such as a public service agency or the legal system. Because the decision maker in each situation is usually an adult, knowledge of how adults learn is important to the application of the Intersystem Model.

Adult Learning Theory

In the nurse-client relationship, the role of the nurse is that of facilitator of the learning, whereas the client is the recipient of the learning. Understanding the expectations that each of these bring to a learning situation is important to helping the nurse be more effective.

The early work on adult learning theory arose out of studies of vocational education by Houle (1961), Tough (1967), and Knowles (1975). Knowles (1980) identified a number of characteristics of adult learners that have formed the core of much of what we understand about adult learning. He characterized the adult learner as one who

1. Is moving toward an independent style of learning
2. Approaches learning with a rich background of experience
3. Is interested in immediate application of the learning

Knowles (1975) and Tough (1967) emphasize the concept of self-determination in learning through active decision making on the part of the learner. They believe that a self-directed learner will be able to diagnose learning needs, set learning objectives, identify resources to use in the learning process, and, finally, evaluate if the learning has produced the desired effect. These concepts have been the core of the adult learning movement in recent years.

Androgogy is the term Knowles (1975) used to describe these principles. He distinguished androgogy from pedagogy, in which the teacher is in charge of the learning process. In androgogy, the subject expert's role has been defined as a facilitator rather than a teacher.

Recently, several adult educators have questioned the conceptual basis for androgogical learning. One problem with androgogy as described by Knowles (1975) and Tough (1967) is a possible cultural difference in learning styles. The majority of studies that have been used to describe the adult learner have been with middle-class, educationally advantaged learners (Brookfield, 1985). Brookfield questions if these principles remain valid with adults from cultures other than the white middle-class culture found in the United States.

The second problem with androgogy is the issue of who is in control of the learning. Much of the literature concerning adult learning has suggested that learning occurs in isolation using tools such as self-study courses, programmed learning, and computer-assisted instruction. Although learning using these tools takes place in an independent setting, the goals and objectives for the learning and the content materials have been developed by a subject expert; thus, the learner is really not in control of the content.

In true independent learning, the learner retains personal power over the learning situation. If a problem is encountered, the learner may consult an expert in the subject area but does not feel compelled to profit from the subject matter's expertise. The freedom to accept or reject the wisdom of the subject expert is one of the hallmarks of true adult learning (Long, 1989).

Long (1986, 1988) views the principles developed by Knowles (1975) and Tough (1967) to be unidimensional. In addition to looking at the responsibilities of the adult learner in directing learning, Long (1989) suggests that a third dimension, the psychological, should also be considered. He believes that there is not uniformity in the ability of adult learners to successfully maintain active control over their learning. The learner who is not psychologically self-directing will not be successful in autonomous learning. He suggests that the degree of success of the learner can be related to the psychological and sociological attributes of the learner. These attributes are culturally determined.

One's culture perhaps has an important influence on the development of a person's learning style. Brookfield (1985) suggests that a person's orientation toward field independence as opposed to field dependence as defined by Witkin (1969) influences the degree of independence a person will have in directing his or her self-learning. His work suggests that a person who grows up in a culture that stresses respect for authority, has clearly defined role definitions, and has an important social control element tends to be field dependent. On the other hand, the person who develops in a culture in which open democracy is promoted is more likely to develop a stronger field-independence orientation.

The constructs used in Long's (1989) theory are the degree of psychological control and the degree of pedagogical control. The independent learner is most successful in a situation in which the pedagogical control is low. The dependent learner, however, does best in the setting in which the pedagogical control is high. When the dependent learner is placed in a low-pedagogical-control setting, successful learning does not occur. These learners do best in a more traditional setting in which the goals and direction of the learning are established by the subject expert. The person with a strong field-independent orientation is more likely to be adept at self-learning.

Bonham (1989) has expanded on the field-dependence/independence characteristics described by Witkin (1969). The first type of learner is the "other-directed learner," for whom dependence is placed on the teacher to provide the structure and content for the learning. The second type of learner benefits the "self-directed-instructional mode," in which the learner takes control of the planning. In the classroom setting, the learner's plans may conflict with those of the teacher. The third type of learner engages in the "self-directed-inquiry mode." This learner prefers to develop a learning plan as the learning moves along. In the classroom setting, this learner will often move off at divergent angles from the plans of the teacher.

Role of the Teacher in the Adult Learning Setting

The role of the teacher with adults who are self-directed learners has had considerable controversy. A question that must be considered is whether any act of learning can be totally self-directed. Because of differences in field dependence of learners, a single approach to teaching is generally not successful. A client who is field independent will respond best when given choices of what and how to learn. A person who is more field dependent will respond better to a more directive approach.

The role that Knowles (1980) has advocated for the educator of adults is that of a facilitator. In the androgogical model, the learner is important in the assessment of learning needs. His schema states that the learner

1. Develops a model of desired behaviors or required competencies
2. Assesses his present level of performance in each
3. Diagnoses gaps between the desired performance and present performance

Then, based on this learning-needs assessment, the learner will develop a plan of action to meet desired outcomes. The question that must be answered is, Is it possible for an adult learner to do these steps independently?

Brookfield (1985) suggests that this is not possible. The learner is not in a position to know what behaviors are necessary to achieve an outcome. In terms of health care, a client may have the general goal of learning to manage newly diagnosed diabetes mellitus. One must question whether this adult learner will know that dietary patterns and blood glucose control mechanisms are the tools that are needed. Thus, an important role for the adult educator is to help the learner discover what one's "real interests are and a pathway to achieve them" (Mezirow, 1985).

Brookfield (1985) suggests that the role of the adult educator is to

1. Provide information
2. Serve as a role model
3. Reinforce learning
4. Be a counselor at times of crisis

He also suggests that the adult educator is to assist with the shifting of paradigms, transform the learner's perspective, and replace meaning.

Mezirow (1994) suggests a similar role for the adult educator. His thesis is related to his concept of perspective transformation. In this process, the adult acquires "a new capability for critical reflection, a new mode of interpreting reality through a heightened awareness of the relevance of context" (p. 228). The adult educator role is to help learners look at the initial premises on which their thought is based and critically examine these paradigms. In other words, the role of the adult educator is to help the learner to think critically. The adult educator helps the self-learner to interpret his or her paradigms in a less distorted way.

Conclusion

An understanding of cognitive, behavioral, and social learning theories, critical thinking theory, and adult learning theory are all necessary for full utilization of the Intersystem Model in clinical practice. Some of these are used as the nurse becomes proficient in the use of the model. Others are used as the nurse carries out interventions identified to assist the client in managing health. This review of learning theory will assist the nurse to incorporate theory into clinical practice.

References

Ames, C., & Archer, J. (1988). Achievement goals in the classroom: Students' learning strategies and motivational processes. *Journal of Educational Psychology, 80,* 260-267.

Bandura, A. L. (1965). Behavioral modification through modeling practices. In L. Kramer & L. Ullman (Eds.), *Research in behavior modification* (pp. 310-340). New York: Holt, Rinehart & Winston.

Bandura, A. L. (1971). *Social learning theory.* Englewood Cliffs, NJ: Prentice Hall.

Bandura, A. L. (1978). The self-system in reciprocal determinism. *American Psychologist, 33,* 344-358.

Bandura, A. L. (1986). *Social foundations of thought and action: A social-cognitive theory.* Englewood Cliffs, NJ: Prentice Hall.

Benner, P. (1984). *From novice to expert: Excellence and power in clinical nursing practice.* Menlo Park, CA: Addison-Wesley.

Bevis, E., & Watson, J. (1989). *Toward a caring curriculum: A new pedagogy for nursing.* New York: National League for Nursing.

Bonham, L. A. (1989). Self-directed orientation toward learning: A learning style. In H. B. Long (Ed.), *Self-directed learning* (pp. 13-42). Tulsa: University of Oklahoma, Oklahoma Research Center.

Brookfield, S. (1985). Self-directed learning: From theory to practice. In G. Darkenwald & A. Knox (Eds.), *New directions for continuing education* (pp. 5-15). San Francisco: Jossey-Bass.

Ennis, R. H. (1985). A logical basis for measuring critical thinking skills. *Educational Leadership, 43,* 44-48.

Facione, N. C., & Facione, P. A. (1992). *The California Critical Thinking Dispositions Inventory.* Millbrae: California Academic Press.

Facione, P. A. (1990). *Critical thinking: A statement of expert consensus for purposes of educational assessment and instruction* (ASHE-ERIC Higher Education Report No. ED 315 423). Washington, DC: Association for the Study of Higher Education.

Gagne, R. M. (1972). Domains of learning. *Interchange, 3,* 1-8.

Gagne, R. M. (1980). Learnable aspects of problem solving. *Educational Psychologist, 15*(2), 84-92.

Gagne, R. M. (1984). Learning outcomes and their effects: Useful categories of human performance. *American Psychologist, 37,* 377-385.

Gagne, R. M. (1985). *The conditions of learning* (4th ed.). New York: Holt, Rinehart & Winston.

Gredler, M. E. (1992). *Learning and instruction: Theory into practice* (2nd ed.). New York: Macmillan.

Houle, C. E. (1961). *The inquiring mind.* Madison: University of Wisconsin Press.

Knowles, M. (1975). *Self-directed learning: A guide for teachers.* Chicago: Association Press.

Knowles, M. (1980). *The modern practice of adult education: From pedagogy to androgogy* (2nd ed.). Chicago: Follett.

Kurfiss, J. G. (1988). *Critical thinking: Theory, research, practice, and possibilities* (ASHE-ERIC Higher Education Report No. 2). Washington, DC: Association for the Study of Higher Education.

Long, H. (1986). Self-direction in learning: Conceptual difficulties. *Lifelong Learning Forum, 3,* 1-2.

Long, H. (1988). Self-directed learning reconsidered. In H. Long & associates, *Self-directed learning: Application & Theory.* Athens: University of Georgia, Adult Education Department.

Long, H. B. (1989). Self-directed learning: Emerging theory and practice. In H. B. Long (Ed.), *Self-directed learning* (pp. 1-11). Tulsa: University of Oklahoma, Oklahoma Research Center.

Mezirow, J. (1985). A critical theory of self-directed learning. In S. Brookfield (Ed.), *Self-directed learning: From theory to practice* (pp. 17-30). San Francisco: Jossey-Bass.

Mezirow, J. (1994). Understanding transformation theory. *Adult Education Quarterly, 44,* 222-244.

Skinner, B. F. (1953). *Science and human behavior.* New York: Macmillan.

Skinner, B. F. (1989). *Recent issues in the analysis of behavior.* Columbus, OH: Merrill.

Tough, A. (1967). *Learning without a teacher* (Educational Research Series No. 3). Toronto: Ontario Institute for Studies in Education.

Watson, G., & Glaser, E. M. (1964). *Watson-Glaser critical thinking appraisal manual.* San Antonio, TX: Psychological Corporation.

Weiner, B. (1980). A cognitive (attribution)- emotion-action model of motivated behavior: An analysis of judgements of help-giving. *Journal of Personality and Social Psychology, 39,* 186-200.

Weiner, B. (1982). The emotional consequences of causal ascriptions. In M. S. Clark & S. T. Fiske (Eds.), *Affect and cognition: The 17th annual Carnegie symposium on cognition* (pp. 185-208). Hillsdale, NJ: Lawrence Erlbaum.

Weiner, B. (1985). *Human motivation.* New York/Berlin: Springer-Verlag.

Witkin, H. A. (1969). Social influences in the development of cognitive style. In D. A. Goslin (Ed.), *Handbook of socialization theory and research.* New York: Rand McNally.

13

Fostering Mutuality
in Clinical Practice

Margaret M. Conger

Interactive Nursing
Teaching the Concept of Mutuality
Patient Vignettes
Conclusion

Nursing has long espoused joint planning with client and nurse for development of a plan of care (Joint Commission of Accreditation of Healthcare Organizations [JCAHO], 1989). In practice, however, this is not commonly done. Many health professionals operate on the premise that they know best what is good for the client. The values of paternalism and beneficence are strongly held beliefs in much of clinical practice. The use of the Intersystem Model in clinical practice provides an excellent opportunity to facilitate nursing students' ability to incorporate the client's value system in care planning. The shared decision making that incorporates both the client and nurse values in the development of a plan of care is defined in this chapter as mutuality.

In this chapter, examples are presented in which student nurses have been effective in identifying client value systems and developing nursing interventions based on mutually agreed-on goals. The students cared for these clients during a clinical rotation in an Advanced Nursing Practicum at Northern

Arizona University. Each example is explored using the Intersystem Model as a framework.

The geographical environment in which these situations occur is important. The client population includes a large American Indian population that still retains many traditional values, along with a geriatric population living in retirement types of communities. In addition, the local medical center serves not only the immediate community but also a large population of tourists who develop health problems as they travel in the high altitude in Northern Arizona. This very diverse client population makes this a rich area for nursing students to develop skill in meeting unique client needs.

Interactive Nursing

A number of authors (Benner & Wrubel, 1989; Gordon, 1987; Orlando, 1972) focus on the importance of an interactive experience between the nurse and the client. Nursing can no longer be "doing for the patient," but rather must involve working with the client to determine mutual goals.

Orlando (1961) identifies the need for the nurse to focus on the individuality of the client. From her early work to the present, she has maintained that the nurse-patient interaction is the most important component of nursing care. Her perception of the nursing process is that the nurse responds to an observed client behavior by sharing his or her understanding of the behavior with the client. The nurse and client then together identify appropriate actions to resolve the problem identified in the behavior. The nurse does not assume that his or her assessment of the situation is correct unless its validity is explored with the client (Orlando, 1972).

Also, the nurse cannot rely on past experiences to know the correct intervention for a particular client. Two patients who present with the same behavior may require interventions that are quite different. Even hospital protocols cannot serve as appropriate reasons for deciding on the use of a particular intervention. At times, it is the responsibility of the nurse to resolve conflicts between the client's need for help and institutional policies (Orlando, 1972).

Benner and Wrubel (1989) also see the need for the nurse to function as the client advocate. A role of the nurse is often to act as a client advocate to the physician and family. Many times, the physician or family concerns differ from those of the client; in such cases, the need for an advocate for the client is essential.

Gordon (1987) explores the need for mutuality between the nurse and client as the plan of care is being developed. She states that the client must contribute personal perceptions of the problem under consideration and plans for health care practices. The nurse must consider the meaning of the situation to the patient. In this encounter, both the patient and nurse value systems must be considered.

Jones and Brown (1993) state that decisions for determining the nursing care required must be made based on the patient's point of view. The nurse and patient need to negotiate between alternate points of view that incorporate the situational context.

Benner and Wrubel (1989) also look at the nurse-client interaction as the plan of care for the client is developed. They state that "if the plan of care does not incorporate input from the patent, a gulf is created between the nurse and patient" (p. 91). They state that the personal concerns of the client are what determine what is "at stake" for a person in any situation. It is the responsibility of the nurse to interpret the concerns that influence the client's understanding of the illness.

The client's culture must be taken into account in the process. Even when the culture of the client is different from that of the nurse, some understanding of culture must be attempted. Benner and Wrubel (1989) suggest that, through elements of shared culture, it is possible to develop an understanding of a culture different from one's own.

Munhall (1993) has described some of the difficulties that are encountered when attempting to see a problem from another's perspective. The nurse needs to spend time communicating with the client, reflecting on what is being said, and then validating his or her understanding of what the client has said. It is hard for the nurse to live out the belief that personal perceptions of the world and of health may not be held by the patient.

Teaching the Concept of Mutuality

Discussion of how to teach nursing students the concept of mutuality is important. Because this teaching is more directed to the affective domain as opposed to the cognitive domain of learning, classroom exercises and discussion alone probably are not sufficient. In my teaching, I introduce this concept in an early lecture class and have the students participate in a case study in which they identify the client's value system prior to planning nursing interventions. Such an exercise, however, probably is not sufficient in itself for students to incorporate the value of mutuality into their practice. Continued reinforcement in the clinical area is needed.

The Intersystem Model is a helpful tool to use with students to encourage incorporation of mutuality into their clinical practice. Artinian (1991) has described the interaction process that occurs when the model is used. During the detector phase of the Intersystem Model, the nurse works with the client to seek information about the state of both his or her internal and external environments. The nurse will assess the client for situational sense of coherence in the three areas. The first area is how comprehensible the situation is to the client. Does he have adequate information to make cognitive sense of the situation. Second, how meaningful is the situation to the client? Does life make sense emotionally? Is the challenge of the situation worth the effort it will take to develop adequate coping skills? Finally, is the situation manageable? Are the resources available adequate for the problem to be managed or does the client have a sense of victimization? The data obtained in the detector phase are further clarified in the selector phase. It is important to understand the value system the client brings to the situation prior to developing outcome goals. Only when the client's value system is recognized can a mutually acceptable plan of care be developed. In the effector phase, the strengths and weaknesses of both the client and the nurse related to the identified situation need to be evaluated to determine the best coping response.

In the clinical practicum, students assess the situational sense of coherence of their clients as part of the care plan. In doing this assessment, the student is required to think about how the individual client is coping with his or her health status. An outcome of this assessment is the identification of the client value systems. Information obtained in this assessment forms the basis for development of a plan of care.

Just the requirement of doing this assessment, however, probably is not sufficient for mutuality to be incorporated into actual practice. I believe that it is also the responsibility of the instructor to provide an environmental climate in which the student feels safe in valuing the client's concerns. As will be seen in several of the vignettes that follow, the institutional environment is often hostile to this concept. An important role the instructor can play is in shielding the student from a hostile environment so that mutuality can be achieved. It is hoped that students exposed to nursing practice in which mutuality is valued will internalize this value. In time, as more nursing students incorporate this value into their practice, changes can be made in the institutional environment as well.

Patient Vignettes

Because the clinical setting is the environment in which the nursing student best learns effective nurse-patient interactions, early exposure to differentiating between institutional values and client values is important. The following vignettes are examples of student interactions with clients in which the client's concerns were identified and the student attempted to include the client in mutual planning.

Two of the vignettes focus on the constraints that the environment can bring in preventing a client from reaching a higher state of health. Two focus on identifying spiritual values held by the client. The last vignette focuses on the challenges that working with a client from the nondominant culture will bring.

The amount of time needed for the nurse and client to clarify values can vary from a brief encounter to a more prolonged one. Some of the vignettes occurred in just a matter of minutes. At other times, it may take much longer for the nurse to be able to identify the client's value system.

The first student vignette describes the interaction of Amy, a senior nursing student, and B, an elderly client with a chronic health problem of diarrhea.

The Wrong Pill[1]

Amy M. Stilley, BSN

"Guess what, it's time to take another pill. Are you ready?" I asked B, a 76-year-old patient who had been lively and cheerful throughout my shift as a team leader in a rural hospital.

"Oh sure, I'm ready. I'm always ready."

"OK, now what pill is this?" I asked, holding a small white pill.

"I don't know. I can't keep track of them anymore because there are so many different kinds."

"Well, this is Lomotil. Do you know what it is for?"

"It's so I won't get diarrhea."

"Right." I then gave him the pill cup and watched as he swallowed the pill.

"You know, that is the generic brand, huh?"

"It sure is," I replied, thinking that it was good that the hospital saved the patient money by giving generic medication whenever possible. I was caught off guard when B's reaction was not what I expected. His face grew very serious and he said,

> I have tried both kinds, the generic and the brand name. The generic just does not work as well as the brand name. I told my doctor that, but he said the only difference is in the fillers. Well, I don't know, but maybe it has something to do with the fillers; but the generic one does not work as well. You know, I am getting old, and I've had surgery on my stomach a few years ago, and have had diarrhea since then. I can't even go to the bathroom normally with this diarrhea. The brand name makes me normal. The generic kind of takes that away, for I'm always trying to make it to the bathroom.

I listened and let him know that I understood. I told him that I would try to get him the brand-name form. I called the pharmacy to explain the situation and found that the brand name was not carried. The pharmacist, however, said that B could bring in his own pills and they would then be supplied to him in place of the generic brand. The doctor would need to write an order to cover this exchange.

I found B's wife and asked her to bring in the medication when she returned to the hospital that afternoon. I also talked with the doctor and got the order for the substitution of the brand name for the generic brand. At this point, I left for a lunch break.

When I returned from lunch, my supervising RN informed me that B was distressed over not having his Lomotil before eating lunch. She said that she tried to explain to him that he had received the medication, but B would not accept her answer. I quickly went to B's room. He looked very worried and had his arms tightly crossed over his chest.

"Hello, Mr. B."

"I don't want to talk about it. You have to do what you have to and that's it."

"What do you mean?" I asked as I sat down next to him.

"I'm old and I don't work like I used to. Some days I don't think living is worth it at all. For the longest time after my stomach surgery, I could not control myself, but the brand name helps. I'm not rich and don't have money to throw around, but I don't mind paying more for those pills. I feel human again when I am on time. Can you understand what I'm saying?"

"Yes, I understand that the brand name works better, and that they make life bearable for you. Mr. B, you have every right to feel the way you do. It must be very difficult for you right now. I know you got the generic brand of Lomotil before lunch today and that you believe that this brand does not help you as much as the name brand. As soon as your wife brings in the prescription bottle, you will be able to begin taking the name brand. Will your wife be here before four thirty?"

"Oh yes, by three o'clock."

"Great, then your next pill at five should be the name brand."

"Thanks for understanding. You're the only one who would listen."

On returning to B's room 15 minutes later, I found him fast asleep.

Analysis

In this vignette, Amy took time to explore B's value system. She learned that B believed that the brand-name form of Lomotil gave him more control over his diarrhea than did the generic brand. She also learned that the use of the brand-name product made him feel that his life was worth living and was worth the extra amount of money that it cost. Amy also had to evaluate her value system of believing that it was good to save money through use of a generic form of a drug.

Amy's interventions focused on the client's value system rather than the institutional system. She took B's concern seriously, giving validity to his situation. By working with the pharmacy, doctors, and the client's wife, she was able to bypass an institutional restraint and provide the desired pill. She kept B informed of the negotiations needed to achieve the joint goal.

Finally, Amy was able to evaluate the effect of her interventions when she went back to check on B and found him able to rest. This was far different from the very anxious person she had seen earlier.

In evaluating this situation in terms of B's situational sense of coherence, the area of comprehensibility was high—B knew all about his reaction to both the generic and the brand-name form of Lomotil. He found meaning for his life in the control of the diarrhea that the brand name gave him. His ability to manage the situation, however, was at risk. He was fighting against an institutional environment that had a very different value system—namely, that the generic form of drug was best. Only as this value system was seen from the client's perspective could this problem be resolved.

The Twelve-Hour Journey[2]

Vince Martinez, RN, BSN

I participated in a unique clinical experience during my senior year as a nursing student at Northern Arizona University. I was assigned to care for Mike, a 40-year-old Caucasian male diagnosed with histiocytic lymphoma of 2 years' duration. He had been treated at a well-respected cancer institute with aggressive therapy, but his body did not respond, and he was told that his condition was terminal. Mike and his family eventually accepted that he would only live 4 to 5 months and that only palliative treatment was recommended.

Mike lived with his wife, Dana, and his 12-year-old stepson in New Hampshire. He also has two grown daughters who are married and have young children of their own. The family mutually decided to travel to the Western states in a motor home so that they could spend some quality time together. Mike was given a thorough physical exam by his doctor and cleared for this trip.

The first week of the trip went without incident. The family made positive gains toward their goal of quality time. Mike began to feel slightly ill with the onset of diarrhea toward the end of the second week. His family hoped it was just stress related to the traveling, but his condition rapidly deteriorated. He developed a fever, generalized edema, blood-tinged stools, and uncontrolled epistaxis. The family had no option but to stop at the nearest hospital.

Mike was admitted to the Intermediate Care Unit through the Emergency Department, diagnosed with pancytopenia. His initial platelet count was 3,000 and only increased to 9,000 following infusion of 30 units of platelets. He also received several units of packed red blood cells. He developed more gastrointestinal bleeding, oral thrush, gross lymphedema, generalized petechiae, severe muscle weakness, severe body pain, and a deeply bruised upper soft palate.

Mike was presented with the option of having multiple invasive diagnostic procedures to fully understand his current crisis. He realized that these procedures would not alter the course of his disease process and that he would eventually die. Mike subsequently refused all of the procedures.

When I started my first day with Mike, I asked the staff nurse what the plan for Mike would be. He answered, "He's going to die here. He's too sick to travel." The physician had spoken candidly to Mike and his family and had suggested that he would probably not live more than 48 hours with the degree of bleeding he was experiencing.

Mike was placed in a reverse isolation room, given liberal pain medication, a soft diet as he could tolerate it, and no further invasive procedures. The reality of impending death left Mike and his family with many sudden concerns and decisions. Mike spoke openly of his desire to die at his peaceful farm home in the presence of his family. At first, this did not seem to be a possibility for him. As the day went on, I noted that Mike was becoming increasingly depressed. He and his family expressed frustration and anger over the difficult situation. His insurance would not cover the cost of a medevac flight home, although it would pay for the equal cost of the invasive procedures and hospitalization.

Mike decided to leave the hospital, even though it meant attempting to drive home in the motor home and dying somewhere along the highway. The stark reality of this decision finally prompted the hospital staff to work with him to develop an alternate plan. The social service worker was able to find a low-cost commercial transport that would get him home in about 12 hours. Mike would fly in a first-class section of the plane with his wife and a sister who was a nurse accompanying him. The family would drive him to Phoenix in the motor home and help him on to the flight. Another family member would meet him in Boston and drive him directly to his home. He would theoretically be in his own bed within 12 hours of being discharged from the hospital. The remainder of the family would drive the motor home back to New Hampshire.

The plan immediately hit many stumbling blocks. Airline personnel denied Mike the ticket when they realized how sick he was. They reluctantly agreed to reinstate the tickets when assured that a nurse would be accompanying him on the flight. The family also realized that all but Mike's wife and sister would be traveling about four days and would probably not see Mike alive again after they said goodbye at the airport.

Preparation for this journey was begun. Mike would be given an infusion of both platelets and packed red cells just prior to transport. He was also given PO pain medication and an antiemetic. We made every effort to accommodate the family needs. The reverse isolation guidelines were adhered to by all hospital personnel, but the family members were not required to wear masks, gowns, and gloves. There was always a family member with him at his bedside.

I noticed that the family was evidencing increased stress during the waiting period. I worked with Mike's sister to help her plan for the flight. We put together a kit of supplies to help manage any emergency that might arise. We talked through all of the potential problems and decided on a plan of action for each. The time for his discharge had to be delayed owing to the onset of a new GI hemorrhage. The physician was very pessimistic about the outcome of this transport because of the new bleeding. He said "He'll probably be dead well before he gets home." He did, however, order more platelets to be given. No platelets were available here and we had to wait for some to be flown up from Phoenix.

Mike left the hospital at five o'clock in the afternoon to begin the 12-hour journey. I woke up at five o'clock the next morning and wondered if Mike had actually made it home to die in his bed with his family at his side. I never found out the ending to this story, but what truly matters most is that Mike was allowed to work at realizing his goal of dying at home. His dignity in the dying process was preserved.

Analysis

In this vignette, Vince was able to look beyond the institutional value of Mike's need for intensive medical care to the greater need for Mike to control his own dying process. He determined to become engaged with Mike as a person and care for his personal values rather than the institutional value system. He was able to see that quality of life for Mike meant taking a big risk—the journey home.

His interventions changed from a cure model to a care model. He was able to assist the family to spend quality time with Mike both while in the hospital and again in undertaking a risky journey. He was able to empower the entire family to set out on the biggest risk they would probably ever have to take. In doing so, however, they were able to satisfy their need to provide Mike with a quality death. In evaluating this situation in terms of the situational sense of coherence of Mike and his family, their level of comprehensibility was high. They knew about his disease and its ultimate consequence. The area in which they needed assistance was in the manageability of the situation. They had to work against institutional values of both the hospital and the airline to take Mike home. The positive support that Vince was able to give them in preparing for this journey gave them the confidence that it could be done. Once given the emotional support they needed, they were able to work with Mike to achieve his most cherished goal. Not once did Vince suggest to them that their initial plan for the journey to the West had been foolish and now they were paying the consequence.

The Need for a Telephone Number[3]

Lisa Winters, SN

I was in the Intensive Care Unit caring for a 73-year-old patient, Frank, who was here from Ireland to visit his daughter. While sightseeing in Northern Arizona, he and his daughter were in a motor vehicle accident that left Frank severely injured. I cared for Frank on the 25th and 26th days of his hospitalization. Frank had little toleration for the nurses providing his care. On my first day with him, the RN I was working with began to orient me to all of the technology surrounding Frank. I asked the RN a question regarding his arterial pressure readings. He looked at me with much frustration and said, "You don't know what the hell you're talking about!"

As a million responses raced through my mind, I blankly looked at him and replied, "You're right, Frank. I am a student and hopefully I will be able to learn from you." I honestly felt very threatened by Frank and didn't know how to handle his behavior.

As I started developing rapport with Frank, he came to trust me more. Upon completing the shift, I said goodbye and told him that I would be returning in the morning.

When I walked into his room the next day, Frank said, "I'm sorry for yelling at you yesterday."

I felt that this was an invitation for conversation. I began to explore his feelings, and Frank shared several concerns. As the conversation progressed, we started to discuss religion. Frank told me that he is Catholic and that he had attended church regularly prior to his hospitalization. I asked Frank how he felt about not being able to worship since he entered the hospital. I also asked if he would like to see a priest while in the hospital.

Frank's response was "Oh yes, but I don't have his number."

I assured Frank that I did know the number for a priest in this community and that I would call him. The solution to this missing aspect in Frank's life was one phone call that I made to a priest who would be able to come and worship with him. This phone call took a mere 60 seconds, yet it changed Frank's perspective of his hospital stay.

Analysis

Lisa recognized Frank's behavior as coming from unmet emotional needs rather than just a hostile patient. By accepting his behavior and working beyond it, she was able to open up communication lines. She identified Frank's frustration with his powerlessness with prolonged hospitalization in a high-technology environment where little attention was being paid to how he felt.

As to Frank's situational sense of coherence, Lisa recognized that Frank was powerless to solve his own spiritual distress. He was in a foreign community far from his home and had no family or friends available to him. He knew what he needed once the conversation focused on his religious practices. He showed that he was highly motivated to accept the spiritual ministry from an unknown priest.

Lisa's interventions focused on allowing Frank to express his concerns about his situation. By careful listening—and astute intuition—Lisa was able to identify his spiritual distress. The interaction took only a few moments, but it made a significant difference in his sense of control over his situation.

The resolution of the problem was extremely simple—one short phone call. In evaluating Frank's response to this very simple intervention, Lisa has made some very cogent comments:

In my opinion, Frank was receiving excellent medical care. However, up to this point, he was not approached holistically, encompassing his spiritual need. By really listening to Frank and getting to the essence of him as an individual, I was able to change a life, spiritually, emotionally, and as a result, physically.

Forty days after admission to the hospital, Frank was finally transferred to a rehabilitation unit. As the nursing staff said goodbye, several of us were needed to carry his belongings to the ambulance. As Frank was wheeled outside for the first time in many weeks, he was beaming. It was a smile I'll never forget. He looked toward the sky and said, "Ah, the sun!" Frank then waved at the staff and went on his way to a truly magnificent recovery.

A Missed Identity[4]

Margaret Bond, RN, BSN

When I first met Robert in a nursing home, he was slumped in his wheelchair and looked at me with sad, melancholic eyes. Tethered to his room by his oxygen tubing, the furthest Robert could venture was into the hallway, just past his room, where he sat quietly in his wheelchair, watching as the nursing home staff hurried about. Robert had been placed in the nursing home 4 months prior, following an episode of congestive heart failure that had left him too weak to ambulate more than a few steps.

During my assessment of Robert, he told me that, although the nursing home record identified his occupation as a retired masonry contractor, he was also a Christian minister, which he considered his "true" profession. He described his current health as "I'm an invalid, I can't do anything I want to do. That depresses me." Robert was a Hopi Indian. He expressed dissatisfaction with having to eat "white man's food" and longed for his traditional foods. When I asked him what I could do to make his life more comfortable, he responded by saying that he wanted to go outside to walk. He also said, "I like to read the Bible, but my eyes are so bad that I can't."

Based on my assessment data, I determined that a major problem for Robert was spiritual distress related to his inability to practice his own faith as he desired because of his immobility and decreased visual acuity. One of my plans was to obtain a larger print Bible which he could read without any visual aid. I also planned to work with the physician and Robert's wife to provide him with the traditional food he desired.

Obtaining a large-print Bible proved to be a great challenge. Each agency I called cited a lack of resources and referred me to another agency. Finally, in desperation, I contacted a friend who worked in a local religious thrift shop and pleaded my case. My friend located a large-print Bible and had it delivered to my home "at no charge." I was thrilled!

The following Tuesday I took the Bible to Robert and explained that it had been donated anonymously. He asked me to read a passage to him and then he read a passage to me. He again asked me to read to him, as the effort seemed too great for him to read aloud. Afterward we sat quietly while the afternoon sun warmed the room. He seemed to be at peace. Promising to return the following Friday, I left the Bible with Robert.

When I returned on Friday, I found Robert's room empty, his bed stripped. I found the charge nurse and questioned her. "Robert died in his sleep last night." My friend had quietly slipped away. His last days were spent sharing readings from his new Bible with his wife and eating traditional foods she had prepared for him. Robert had reconnected with his spiritual side.

Analysis

In Margaret's assessment data collection process, she found what any nurse could have seen had they stopped to look at Robert. He was the picture of social isolation resulting from his immobility and dependence on the oxygen generated from the compressor in his room. She went a step farther, however, and discovered the unexpected—a Hopi Indian who was a Christian minister. Robert made it very clear to her that he considered his role as a minister to be the important part of his life's work. Robert was also depressed about his inability to do anything he wanted to do. He could not walk outside, and he had to eat the wrong kind of food.

Margaret was able to meet his spiritual need by procuring the large-print Bible for him. She was also able to meet his desire for his traditional food. She tried to get a portable oxygen system but was not successful in working through the bureaucracy for that. The simple intervention of providing for his spiritual need, however, had a life-altering effect on him. Even though he had the Bible only for a few days, it provided him with a peaceful death.

Robert's situational sense of coherence at the time Margaret first met him was very low. He had no ability to manage his life to accomplish even his simple desires related to spiritual practices, traditional food, and ability to go outside. He seemed to have no ability to know how to even begin to procure these things he desired. His motivation was also low, probably as the result of 4 months of living in the nursing home where no one seemed to know or care anything about his desires.

The intervention Margaret used was simple yet very effective. Often, it does not take a great deal of energy to help a client satisfy his spiritual needs. A simple act of caring enough to find out what is important to the client proved to be sufficient.

Who Decides[5]

Elena Kirschner, RN, BSN

The provision of nursing care that is congruent with perceived client desires can be an extreme challenge. This was my experience in caring for an 89-year-old Navajo grandfather and traditional medicine man. The client had been transferred from a small hospital on the reservation for surgical repair of a yet unset fractured hip. Once admitted to the Intermediate Care Unit, he was given a long list of medical diagnoses: fractured hip, end-stage renal disease (ESRD), hypercalcemia, anemia, ruptured tympanic membranes, sepsis, and malnutrition. The associated high-technology interventions for each medical diagnosis were immediately initiated.

It was apparent to me, as a nursing student, that the value system of the hospital unit and its staff, rather than the client's wishes, were guiding the course of care for this patient. The primary medical goal was stabilization of the patient for surgery. This remained the goal for several days after it was clear that the patient was experiencing multisystem failure and was unlikely to survive his hospital stay.

To achieve this stabilization, the client's nutritional status needed improvement; therefore, an attempt was made to place a central line for the delivery of total parenteral nutrition. The client fought this procedure, and the attempt was unsuccessful. He was left with gross ecchymoses from shoulder to groin. Even though the first attempt at placement of the central line was extremely difficult for the client, the staff nurse, while obtaining a telephone consent from the family for a second attempt, described the procedure as "strictly routine." This attitude is an example of the institutional value system that persisted despite the nonverbal objections of the client.

Lack of a common language and client deafness made verbal communication impossible. Thus, nursing assessments had to be based on careful observation of the client's responses to interventions. A good understanding of the cultural context of the situation was also helpful. It was my observation that the client strongly objected to invasive procedures, especially those requiring movement. He objected by using hand movements motioning away the staff. He also used facial grimacing, loud and disapproving speech, and agitation. As more and more procedures were performed, the client's disapproval became less overt. He began to withdraw into a semi-conscious state. Objective findings such as an increased blood pressure, tachypnea, tachycardia, and diaphoresis during procedures suggested a pain response.

For me, the client's withdrawal indicated that the situation (continuous invasive procedures) was becoming intolerable to him. I identified the nursing diagnosis of pain and social isolation and began interventions related to them. My nursing interventions included regard for the client's need for contact. Regular hand holding and gentle touching were instituted. I made contact with a Navajo translator, and daily social contacts with a Navajo speaker were arranged. I also advocated for a pain management protocol. I eventually was able to get the attending physician to prescribe the pain medication needed.

The client's response indicated that these interventions were well received. He seemed to regain a sense of control when his admonitions affected staff behavior (his complaints resulted in removal of painful stimuli). He was greatly calmed by touch and reached out for a hand when offered. He was able to rest more easily and endure procedures when given the pain medication on a regular basis.

The client's interest and attention increased when he was spoken to in his own language. The Navajo translator was able to verify my understanding of the cultural response to the high technology on the part of this traditional medicine man. She indicated that the cultural value was a harmony with nature and that high-technology interventions that prolong suffering were probably in conflict with the client's belief system.

The medical prognosis for this client was extremely poor. As his nurse and advocate, I felt that his nonverbal communication efforts should be heeded and he be allowed a comfortable and dignified dying process. However, this was not the institutional value system and invasive procedures continued. These included placement of the central line, TPN administration, daily hemodialysis, IV antibiotic therapy, invasive suctioning, and a host of miscellaneous procedures.

I attempted to provide nursing care that honored the client's desires. I was only partially successful in this goal. My status as a nursing student was prohibitive to effecting a change in the existing mindset. Therefore, comfort care was unfortunately provided, so to speak, between the cracks. Today, as a staff nurse, I use the frustration of this situation to remind me to base my nursing practice on the needs of the client and not the institutional setting.

Analysis

Elena used data sources for assessments that were in accord with the client situation. Because of the language barrier and the deafness of the client, she focused on the nonverbal clues given. She also used her understanding of Navajo culture to predict the response of an elderly medicine man to the high-technology environment of the modern hospital. Her understanding of the culture was further validated by using the resources of a Navajo translator.

Although Elena's interventions were not successful in preventing further painful interventions for this client, she was able to reduce his discomfort by providing for pain needs and giving emotional support through touch. Perhaps the most important outcome of this interaction was the learning provided to the student. Elena has pointed out how this experience has affected her practice as a nurse. She now is in a position in which she can have a greater effect on the institutional practices and has been able to better advocate for clients to prevent the use of high-technology approaches that are inconsistent with the client's belief system.

The Navajo medicine man's situational sense of coherence at the time Elena cared for him was extremely low. His ability to have any control or management of the situation was not present. He was not consulted about his wishes regarding any of the medical treatments. His knowledge level also was low. He had had little prior contact with the high-tech hospital and had no family members present to intervene for him. His motivation to participate in the therapies presented was low. During the time Elena cared for this client, she was able to begin to give him a little control over the situation. Some of the hospital staff began to respond to his nonverbal messages and not insist on performing painful procedures that he did not want. After several days of this negotiation, the client was moved out of an intensive care environment and was allowed to die peacefully.

Conclusion

The interventions presented in this chapter were guided by the student's evaluations of issues of profound importance to each of the clients. In each case, the concepts outlined by Orlando (1972) were employed. It was necessary to clarify with the client what his or her true needs were. Even in the case of Elena working with the Navajo grandfather who was unable to communicate in English, the client's value system was identified through observing his body language. His response to comfort measures was seen as approval of the interventions. As the uniqueness of each of these clients was validated, and as the plan of care was altered to include the value system of each, the deepest needs of the clients were met and each increased in situational sense of coherence. This occurred despite that in three of the five vignettes, the client's final outcome was death.

Each of the interventions was focused at the deepest level of the person's being. The nurse made no attempt to try to bring the client into conformity with the institutional values present. It would not have been appropriate for the nurse to try to alter the client's value system. Instead, he or she worked to bring the institutional value system into accord with that of the client. In this way, the client advocate role was expressed and the nurse was able to meet the identified need.

According to Orlando's (1972) conceptual framework, if the client's need has not been clarified, then the nurse has done nothing. In each of these vignettes, the student was able to identify and meet the client's most fundamental needs by moving beyond the ordinary practice of nursing care. A truly individualized plan of care in which the client and nurse were mutually active in the goal-setting process was accomplished. These vignettes demonstrate that mutuality is possible in today's health care environment.

Notes

1. Used with permission from Amy M. Stilley.
2. Used with permission from Vince Martinez.
3. Used with permission from Lisa Winters.
4. Used with permission from Margaret Bond.
5. Used with permission from Elena Kirschner.

References

Artinian, B. (1991). The development of the Intersystem Model. *Journal of Advanced Nursing, 16,* 194-205.

Benner, P., & Wrubel, J. (1989). *The primacy of caring.* Menlo Park, CA: Addison-Wesley.

Gordon, M. (1987). The nurse as a thinking practitioner. In K. Hannah, M. Reimer, W. C. Mills, & S. Letourneau (Eds.), *Clinical judgement and decision making: The future with nursing diagnosis.* New York: John Wiley.

Joint Commission of Accreditation of Healthcare Organizations. (1989). *Nursing standards task force: Revision process.* Chicago: Author.

Jones, S. A., & Brown, L. N. (1993). Alternative views on defining critical thinking through the nursing process. *Holistic Nurse Practioner, 7,* 71-75.

Munhall, P. (1993). "Unknowing": Toward another pattern of knowing in nursing. *Nursing Outlook, 41,* 125-128.

Orlando, I. J. (1961). *The dynamic nurse-patient relationship: Function, process, and principles.* New York: G. P. Putnam.

Orlando, I. J. (1972). *The discipline and teaching of nursing process.* New York: G. P. Putnam.

14

Delegation Decision Making

Nursing Assessment Decision Grid

Margaret M. Conger

The ability of registered nurses (RNs) to make decisions about the delegation of nursing care to lesser-skilled nurses has become an important component of nursing practice as an increased number of non-RNs have been introduced into the acute care hospital. Educational programs or clinical experience, however, have not prepared staff nurses to function in the delegation decision role. To make delegation decisions effectively, the RN must learn new skills.

Conceptual Framework for the Development of the Nursing Assessment Decision Grid

The Nursing Assessment Decision Grid (NADG) shown in Figure 14.1 is a tool developed for use in teaching RNs to identify those aspects of nursing

DELEGATION
WHEN IS IT SAFE?
TO WHOM CAN I DELEGATE?
QUESTIONS TO ASK YOURSELF

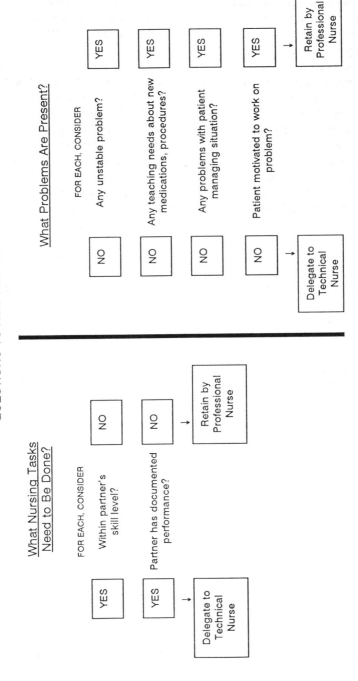

Figure 14.1. Nursing Assessment Decision Grid

care that they must retain and those aspects that can safely be delegated to a lesser-skilled nurse (Conger, 1994). Its design is such that it leads the learner through a number of steps to analyze a patient situation and then make a decision about the nursing needs of that patient. These steps have been developed using the Intersystem Model as the conceptual framework.

The Nursing Assessment Decision Grid requires the learner to implement three steps (see Figure 14.2) using an assessment worksheet. In the first step, the learner lists the tasks that he or she identifies that are required to provide patient care. These tasks are pieces of assigned work. In nursing, these tasks are generally the outcome of physician orders (e.g., administration of medications or treatments) or nursing orders (e.g., the need to turn a patient frequently).

The second step is to consider each of the biological, psychosocial, and spiritual subsystems as identified in the Intersystem Model (Artinian, 1991) to evaluate health problems that a patient may be experiencing. For ease in use of the Nursing Assessment Decision Grid in the clinical setting, the spiritual subsystem assessment was grouped with the psychosocial subsystem assessment.

The third step is to evaluate the patient's adaptation to his or her situation using the construct of "situational sense of coherence" (SSOC). This construct was adapted from Antonovsky (1987) and incorporated into the Intersystem Model (Artinian, 1991).

The elements embedded in the situational sense of coherence can be defined as a sense of comprehension or knowledge needed to understand the situation, the ability to manage the situation, and motivation and meaningfulness to be involved in the situation (Artinian, 1991). Because the goal of nursing is to assist a patient to move toward the health end of a wellness-illness continuum, attention to the patient's adaptation to the given situation is important. As the patient's situational sense of coherence improves, movement toward health on the continuum can be seen.

Clinical Use of the Nursing Assessment Decision Grid

The Nursing Assessment Decision Grid (NADG) consists of a set of "rules," or guides, that the nurse uses in the client assessment process. In the classroom setting, in which the nurse is learning to make delegation decisions, the NADG is used in conjunction with patient vignettes developed by Conger (1994). In the clinical setting, it is applied to actual patient situations.

When used in the clinical setting, the nurse collects initial assessment data using the framework of the NADG. Nursing task identification is based on principles outlined by a number of authors (Manthey, 1989; Neuman, 1989;

Step 1
  A. List, on the assessment work sheet, all tasks required.
  B. Rate each task using the following scale:
    1. LVN can do independently
    2. LVN needs to be taught
    3. RN must do

Step 2
  A. Consider each of the client's subsystems listed below. List on your assignment sheet any
     that are causing stress (imbalance) at this time.
     Biological subsystem problems:
     1. regulatory
        a. neurological
        b. endocrine
        c. immune
     2. circulatory-respiratory
     3. musculoskeletal
     4. gastrointestinal
     5. renal
     6. reproductive
     Psychosocial and spiritual subsystem problems
     1. anxiety
     2. family process
     3. fear
     4. grieving
     5. hopelessness
     6. spiritual distress
  B. Rating scale: For each problem identified, rate the rapidity of change occurring at this
     time using the following scale:
     1. stabilized
     2. showing moderate change
     3. rapidly changing

Step 3
  A. Consider the patient's and family's sense of coherence with the present situation in terms
     of knowledge, ability to manage the situation, and motivation to cope with the situation.
  B. Rating scale:
     1. Knowledge: 1 — has adequate knowledge
                          2 — needs some instruction/review
                          3 — needs intense instruction
     2. Manageability   1 —able to manage the problem
                          2 — needs minimal assistance to manage the problem
                          3 — needs extensive assistance to manage the problem
     3. Motivation   1 — not motivated to work on problem at this time
                          2 — shows moderate motivation to work on problem
                          3 — highly motivated to work on problem

Figure 14.2. Steps for Using Nursing Assessment Decision Grid

Poteet, 1984; Rabinow, 1982). Many nursing tasks fall into the category of mandatory practice because the source of these assignments is physician orders or institutional policies. Other tasks are mandated by professional practice through nursing orders. Patient problem identification is more complex.

The conceptual framework for problem analysis in the NADG is the construct of person as defined in the Intersystem Model (Artinian, 1991). In this model, patient problems arise from subsystems within the person. The spiritual system is seen at the center of person's being and is made up of the subsystems of beliefs, values, religious practices, and affective responses. This subsystem is surrounded by the psychosocial system. The subsystems in the psychosocial dimension include the cognitive, the affective, and the interactional. The outermost subsystem is the biological. It is composed of the following subsystems: (a) regulatory, which includes the neurological, endocrine, and immune; (b) circulatory and respiratory; (c) musculoskeletal; (d) gastrointestinal; (e) renal; and (f) reproductive.

In addition to assessment of the identified patient subsystems, an assessment of the knowledge base, the values, and the ability of the patient to manage personal care needs is seen as pertinent to patient problem identification. Both positive and negative factors that affect the patient are analyzed using a scale that rates the individual on the situational sense of coherence (Artinian, 1991). This construct includes not only the components of a person's ability to comprehend and to manage a situation but also the person's assessment of the meaningfulness of the suggested action to health care outcomes or personal well-being. Using these components to evaluate a patient's response to a situation, the nurse and patient together then formulate a plan of care that would increase the patient's SSOC and allow movement toward the health end of a wellness-illness continuum.

Nursing Assessment Decision Grid Scoring Strategies

Scoring on Nursing Tasks

Both the tasks that are required to care for the client and the nursing problems that are identified through a review of the client's biological, psychosocial, and spiritual subsystems are analyzed using the rating scale found in the NADG. The scoring of each of these databases is based on a 3-point scale.

Scoring of nursing tasks is based on education and demonstrated competence of the staff member, licensing laws, and institution policy relative to nursing practice. Those tasks that are restricted to RNs by law, such as the administration of intravenous medications, are always rated as a 3. Those tasks that a licensed vocational nurse (LVN) can do according to institution policy are rated as either 2 or 1. If the LVN has not yet demonstrated competence in the task, it would be appropriate to teach him or her at this time, giving the task a rating of 2. If the LVN has already demonstrated competency in the task as demonstrated by a work-related skills assessment program, this task should be rated as 1 and handled independently by the LVN.

Scoring on Patient Problems

Rapidity of change in a patient problem is one parameter that can be used in determining if the care can be delegated to an LVN. If a patient is experiencing a problem that is showing rapid deterioration—for example, a quick decline in respiratory function—the close assessment of a professional nurse is required, and a score of 3 would be assigned. If the patient's response to a problem is stabilized (e.g., a patient with chronic respiratory disease in which the frequency and quality of the assessment needed would be less than that of a patient with an acute problem), the care of this patient could be safely delegated to an LVN. This situation would be assigned a score of 1. In contrast, a patient who has a problem that is gradually showing deterioration would need more frequent assessments. The RN, however, could give guidance to an LVN to observe for specific parameters and thus delegate the care of the patient to him or her. This situation would be assigned a score of 2.

Scoring on Situational Sense of Coherence

The scoring on the situational sense of coherence follows the general guideline established by Artinian (1991). The RN should first assess the patient's motivation to participate in his or her care. If a patient shows a high degree of motivation (rated as 3), the RN needs to retain the care of this patient because a great deal of emphasis will be placed on patient and family teaching. These are responsibilities that have been determined to be within the scope of the RN by both nursing ethics and law (American Nurses Association, 1985; California Nursing Practice Act, 1988).

If the patient does not show motivation to participate in his or her care, he or she is rated 1, and the care can be assigned to an LVN. If the level of motivation is assessed as moderate, the nurse should rate this situation as 2. In such a case, the care could be assigned to either the RN or the LVN. The assignment decision should be based on the total demands of the day's assignment.

The knowledge and manageability needs of patients are also determined in a similar manner. A knowledge need that is new to the patient should be rated 3 because it would require the skill of an RN to assess his or her learning needs and set up an appropriate teaching program. If the patient simply needs reinforcement of prior learning knowledge need should be rated 1. An example of this would be a patient who has been following a salt-restricted diet for a period of time; knowledge about such a diet should be rated 1 because reinforcement of the principles of this diet would be within the legal functions of the LVN. A patient needing moderate teaching would be rated as a 2 and the assignment decision would be based on the total demands of the day's assignment for the RN and LVN. When the patient is well aware of diet restrictions but continues to select inappropriate foods for consumption, a 3 would be assigned because the RN will need to assess what barriers contribute to this continued self- destructive behavior.

The scoring for the patient's ability to manage health needs would be similar. A patient with a problem that will require considerable assistance for management (e.g., a new colostomy) requires the intervention of an RN and would be scored as a 3. A patient with a problem of lesser severity (e.g., crutch walking) could be handled by an LVN and scored as a 2. A problem that the patient is able to manage independently or requires only minimal assistance would be scored as a 1.

Educational Preparation for Use
of the Nursing Assessment Decision Grid

Nurses' skill in identification of both nursing tasks and patient problems needs to be developed. In the clinical setting, the nurse needs to be able to rapidly identify the critical elements of the care required for a client on his or her hospital unit and make appropriate decisions. Educational programs utilizing the NADG are intended to achieve this outcome.

The effectiveness of the NADG in improving the delegation decision making of RNs has been tested in the classroom setting (Conger, 1994). Patient vignettes depicting actual clinical situations were developed by the researcher. Classroom activities were then planned in which the participants were first introduced to the NADG and then led through an exercise in which the nursing tasks and patient problems found in a vignette were identified. The participants then used the rating scale to make decisions about which aspects of care needed to be retained by an RN and which could be delegated to an LVN partner.

The teaching-learning environment established in the classroom setting was based on the 4MAT System as developed by McCarthy (1981) (see Figure 14.3). In this environment, differences in individual learning styles are recognized and instructional techniques are used that will encourage learning in all types of learners. The inclusion of learning activities to meet all types of

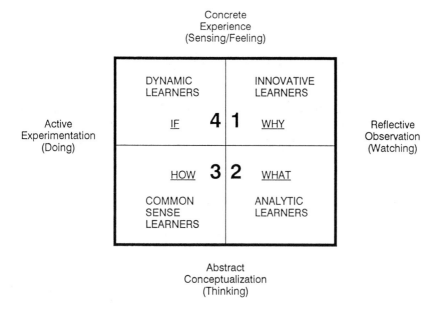

Figure 14.3. Learning Styles Used in 4MAT Design (McCarthy, 1981).

learning needs is especially important in a class setting in which the instructor is unfamiliar with individual participant learning needs.

McCarthy (1981) has categorized learners into four styles by the nature of the question they ask of a learning experience. The innovative or Type 1 learner is interested in the personal meaning of the lesson. The preferred method for learning is through the use of sensing and feeling. This learner is described as the person who asks the question "Why." When working with this style of learner, a clear objective for the lesson must be established early or the learner will "tune out" what is happening and not participate in the activity.

The analytic or Type 2 learner is characterized by an interest in facts. The question asked by this style of learner is "What." This learner needs to be able to develop new concepts into an organized system to deal with them effectively.

The commonsense or Type 3 learner is characterized by an interest in how things work. This style of learner works best by actively trying out a new idea to find out how it works. This process allows the person a means of becoming actively involved in the learning experience.

The dynamic or Type 4 learner is characterized by an interest in self-discovery. This style of learner frequently uses the question "if." The individual wants to expand on the learning and extend it to new situations.

Classroom activities were designed to incorporate each of these styles of learning so that the effectiveness of the NADG would not be biased by a classroom environment not conducive to some of the participant's learning.

The theoretical construct used for the patient vignettes used in the analysis process is based on the novice to expert concept developed by Benner (1984). Benner suggests that an effective learning strategy for novices is the use of vignettes with a set of "rules" to guide the learner in the decision-making process. She also suggests that repeated use of a problem-solving approach will lead to skill acquisition that can be transferred to the clinical setting.

Outcome Assessment for the Nursing Assessment Decision Grid

Staff nurses working in an institution in which LVNs were scheduled to be brought into the workforce were required to attend a class in which the delegation principles found in the NADG were presented. A pretest-posttest design was used to study the effectiveness of the NADG in improving these nurses' ability to make delegation decisions.

The pretest consisted of two patient vignettes in which the participants were asked to identify all the nursing tasks and patient problems and then to identify the appropriate caregiver for the client presented in each vignette. Following the classroom activities in which the principles used in the NADG were presented, the participants were given the same two patient vignettes as found in the pretest to analyze. Changes in the number of correctly identified nursing tasks and patient problems and delegation decision choices were measured.

A complete description of the testing process is described in Conger (1994). The desired outcome of using the NADG in the classroom situation to bring the learner to a competency stage in delegation decision making was met. The RNs participating in the class showed significant improvement in their ability to identify and rate nursing tasks and nursing problems and make decisions about what aspects of care could be safely delegated to a LVN.

One of the gratifying outcomes of this study was that when the participants were divided into high- and low-performing groups based on results of the pretesting, the low-performing group demonstrated a statistically significant gain on posttest compared to the high-performing group. This finding suggests that a rule-based educational activity can improve the delegation decision-making ability of nurses shown to be at risk in this area.

Follow-Up Activities

As the nurse moves from the classroom setting to the clinical area, the principles of the NADG need to be reinforced. The chart shown in Figure 14.1 was developed by the author to provide such reinforcement. The principles of the NADG are summarized here. Both wall charts and pocket cards of this material were prepared to provide the nurse with a ready reminder of the classroom learning.

Conclusion

As partnership models of nursing delivery systems increase, teaching strategies to prepare nurses for this new environment are vital. A critical element of any partnership delivery system is the ability of the RN to make safe and cost-effective decisions about the utilization of all the various levels of nurses being introduced into the workforce. It is suggested that use of the NADG is one approach to meet this need.

The conceptual models used in the development of the NADG represent a theory-based approach to practice. The decision process for analyzing nursing tasks was developed from a variety of sources in the nursing literature. The decision process for analyzing patient problems was developed using the Intersystem Model (Artinian, 1991). The teaching strategy used is based on the "novice to expert" concept developed by Benner (1984).

Nursing partnerships between RNs and various other levels of nurses are on the rise. Nursing administrators need to carefully consider the professional implications of such practice. The use of the NADG provides a means for the RN to retain a professional role of client assessment as she or he directs the activities of nonprofessional nursing assistants. Because this tool is grounded in theoretical concepts, it provides a means for transferring theory into clinical practice.

References

American Nurses Association (1985). *Code for nurses with interpretative statements.* Kansas City, MO: Author.

Antonovsky, A. (1987). *Unraveling the mystery of health: How people manage stress and stay well.* San Francisco: Jossey-Bass.

Artinian, B. (1991). The development of the Intersystem Model. *Journal of Advanced Nursing, 16,* 194-205.

Benner, P. (1984). *From novice to expert.* Menlo Park, CA: Addison-Wesley.

California Nursing Practice Act, Cal. Health & Safety Code § 2758 (1988).

Conger, M. M. (1994). The Nursing Assessment Decision Grid: Tool for delegation decision. *Journal of Continuing Education in Nursing, 25,* 21-27.

Manthey, M. (1989). The role of the LPN or the problem of two levels. *Nursing Management, 20,* 26-28.

McCarthy, B. (1981). *The 4MAT system.* Barrington, IL: Excel.

Neuman, T. A. (1989, September). A nurse's guide to fail-safe delegating. *Nursing 89,* 63-64.

Poteet, G. W. (1984, September). Delegation strategies: A must for the nurse executive. *Journal of Nursing Administration,* pp. 18-21.

Rabinow, J. (1982, September/October). Delegating safely within the law. *Nursing Life,* pp. 48-49.

15

Research

Refining and Testing the Theoretical
Constructs of the Intersystem Model

Barbara M. Artinian

Management of Chronic Obstructive Pulmonary Disease
Communication With the Ventilator-Dependent Patients
Risking Involvement With Cancer Patients
The Process of Regaining Control
Strategies for Increasing Meaningfulness
The Effect of Congruency of Expectations
Discussion

The Intersystem Model provides a framework for implementing nursing practice and guiding research. Because the model was developed within the perspective of symbolic interaction theory, that framework "sets the selection and formulation of problems, the determination of what are data, and the means to be used in getting data, the kinds of relations sought between data, and the forms in which the propositions are framed" (Blumer, 1969, p. 25). The model also directs the researcher to the types of problems that are relevant.

Symbolic interaction theory, as developed by a group of sociologists at the University of Chicago known as the "Chicago School," focuses on the nature of human social interaction. Therefore, research methodologies that are appropriate for use with the Intersystem Model are those that focus on the nature of the shared meanings with which individuals and groups create symbolic worlds. Qualitative research, with its interpretive, naturalistic approach to the world, is especially appropriate because it allows the study of things in their natural setting and the interpretation of phenomena in terms of the meanings people bring to them. Research materials include case studies, personal experience, interviews with individuals or groups, observation of interactions, and collected documents that describe the phenomenon of interest. The goal is to take the role of the other to understand the phenomenon from the perspective of the subject.

Because the focus of symbolic interactionism is on "how people define events or reality and how they act in relation to their beliefs" (Chenitz & Swanson, 1986, p. 4), the major concern of research within this tradition is to identify the meanings that events hold for individuals and others with whom they share experiences as they construct their lives together within a particular social context. The social reality of the subjects can be understood by exploring such issues as (a) the meanings of relationships to them and the power dynamics underlying those relationships; (b) the process of decision making; (c) the consequences of social support; (d) the empowerment of patients through the provision of clinical knowledge that allows them to understand their own realities; (e) an understanding of the lifestyle, values, beliefs, hopes, and commitments of the subjects; (f) the making sense of strategies used by subjects to manage crises within their personal context of motivations and preferred outcomes; (g) the interactional demands of a particular lived experience; and (h) the interconnections between health care providers and patients or clients (Tourigny, 1994).

The research method of choice is qualitative research that allows extended periods of data gathering within the social context of the subject's experience. Qualitative methodology pays attention to the importance of the relationships between the researcher and subjects in determining what is seen and revealed and, therefore, what is more appropriate for understanding intersubjective meanings. Tourigny (1994) discusses the value of qualitative research that allows the subjects to express themselves within the context of their lived experience:

> Methods that share an openness to participants playing directive roles in selecting, structuring, and sequencing data they deem most representative of their stories . . . seem singularly suited to the task of empowerment. Because they retain primacy of voice, structure, sequence, and syntax in each narrative, the research itself becomes an opportunity for self-clarification (Ronai, 1992). Respondents hear their own stories within contexts they had not necessarily previously acknowledged as pivotal determinants of their lived experiences. (p. 181)

Grounded theory is a research method derived from the symbolic interactionist perspective. Analysis of data within the symbolic interaction tradition takes into account the interactional, social, and structural aspects of the research experience. It is a methodology for generating theory that is grounded in data that are systematically gathered and analyzed using the constant comparative method (Glaser & Strauss, 1969). In this method, theory can be generated directly from the data or may be expanded and elaborated from existing theories that are relevant to the phenomenon under investigation (Artinian, 1986). The outcome of a grounded theory study can be the identification of abstract concepts and propositions about the relationships among them (Artinian, 1982). Chenitz and Swanson (1986) state that grounded theory studies can be reported at both the descriptive category level and the theoretical or process level.

In an attempt to further develop and expand the theories and constructs incorporated into the Intersystem Model, a number of research studies have been proposed and conducted within the symbolic interaction tradition. Research studies designed to study aspects of the Intersystem Model include qualitative descriptive studies (Hecox, 1996; Stilwell, 1995), grounded theory studies (Artinian, 1995a, 1995b, 1996b; Cone, 1994), studies combining quantitative and qualitative approaches (Artinian, 1996a), and the development and testing of an instrument that incorporated the components of the situational sense of coherence (SSOC) construct to measure the delegation process (Conger, 1991). Each of these studies is described in this chapter or in another chapter of the book.

The development and testing of an instrument to measure SSOC is described in Chapter 2. The delegation instrument developed by Conger (1994) is described in Chapter 14 in relationship to its use in teaching nurses the delegation process. A qualitative grounded theory study, "Connecting: The Experience of Giving Spiritual Care" by Pamela Cone, is presented in Chapter 16.

Understanding How Patients With Chronic Obstructive Pulmonary Disease Manage Their Illness[1]

Judith R. Milligan-Hecox

The physical and emotional disabilities resulting from chronic obstructive pulmonary disease (COPD) can severely affect the patient's quality of life. One who suffers from this illness may have to leave the workforce and curtail social interactions. As independence gradually diminishes, family structure and roles change. With this change also comes an ever-changing view of self. New ways to manage life must be found and with these new ways, new meanings to life.

The purpose of this study was to develop an understanding of how patients manage the changes of COPD, while still retaining a sense of self and whatever independence they can preserve. By uncovering and tapping into the wisdom and understandings of these patients and their family members, the health care professional can share in these strategies and assist others still learning to maintain their independence in the face of COPD. The questions which were explored in this study were the following:

1. What is the effect of COPD on individual and family functioning? How has this changed over time?
2. What do the patient and his family know about managing the illness? What strategies have they developed to manage the illness?
3. How motivated are the patient and family to engage in management of the illness?
4. What use do the patient or family members make of available or potential resources, both material and interpersonal?
5. What benefits do patients and families attribute to the pulmonary rehabilitation program in managing their illness?
6. How is quality of life maintained in the face of an illness that makes breathing itself uncomfortable?

This study was conducted within the framework of symbolic interaction. This framework has as its central guiding force the position that the meanings things have for human beings are central in their own right (Blumer, 1969). Role theory is a sub-theory within the symbolic interaction orientation and encompasses concepts such as role expectation, stress, strain, conflict, incongruence, ambiguity, overload, and negotiation. These concepts were explored to observe how they affect meanings assigned to the illness situation.

Open-ended interviews were conducted with patients enrolled in the Pulmonary Rehabilitation Program at Barlow Respiratory Hospital, both in the rehabilitation setting and at home after discharge from the program. The patients and family members were encouraged to introduce new ideas to provide a clear picture of the experience from their perspectives. The interviews were transcribed and analyzed using the comparative analysis techniques for qualitative data as developed by Glaser (1978) to aid in understanding the strategies used by patients and their families in living with the illness. Major themes that emerged from the data were pacing, depending on others, clarifying values, maintaining independence, maintaining the struggle, and accepting ambiguity.

Communication With the Ventilator-Dependent Patient[2]

Karen Stilwell

A study designed to look at the communication process between the nurse and the ventilator-dependent patient is being developed by Karen Stilwell, a graduate student at Azusa Pacific University. She is interested in identifying the strategies used by the ventilator-dependent patient and the nurse to facilitate communication. Because the ventilator-dependent patient is unable to speak, many adaptive strategies are developed by both the patient and the nurse to facilitate the communication of ideas. Communication breakdowns frequently occur, however, resulting in frustration for both patient and nurse. The aim of her study is to explore the typical interactions between the ventilator-dependent patient and nurses within the context of the hospital environment. The characteristics of the process will be studied to determine if there are specific patterns that indicate communication breakdown.

The questions to be explored in this study are the following:

1. How does the nurse know when further efforts at communication in a given interaction will be ineffective? How is the situation managed?
2. When is it perceived to be worth the effort to continue?
3. What are the characteristics of the breakdown for both the nurse and the patient?
4. How does motivation affect the process, and what are the characteristics of both motivated and nonmotivated communicators?
5. How does the interaction progress?
6. What is the patient's perceived functional ability? How does the self-image affect the interactions?
7. What does effective communication look like?

Participant observation and in-depth interviews with nurses and with patients after they have been freed from the ventilator will be the primary sources of data. Analysis of data will be done using the constant comparative methodology of grounded theory.

Risking Involvement With Cancer Patients[3]

Barbara M. Artinian

The helping relationship has been identified by Brammer (1988) as central to the practice of nursing. This study explored the process of developing special relationships with cancer patients. Interviews were conducted with 32 nursing personnel to investigate the difference between the usual relationships they had with patients in general and the special relationships they formed with selected patients. The following questions were explored.

1. Under what conditions do special nurse-patient relationships form?
2. What are the activities that characterize a special relationship? How are these different from the activities of other relationships?
3. What strategies do nurses use to foster or limit special relationships?
4. What are the consequences of special relationships?

Using the grounded theory method of constant comparative analysis, data were simultaneously collected and analyzed to identify the core process. The basic social process of risking involvement was discovered. It was found to range from the minimal involvement of neutrality to the intense involvement of a "special relationship." These findings are summarized in a conceptual map (see Figure 15.1).

The Process of Regaining Control:
Outcome of a Pulmonary Rehabilitation
Program for Patients With Advanced COPD[4]

Barbara M. Artinian

Chronic obstructive pulmonary disease (COPD) is the second leading cause of disability in the United States. It often results in extreme debility or the need for mechanical ventilation. The social isolation and loss of control that follow often cause the patient to give up hope for independent living. The purpose of this study was to observe the process through which patients moved from the dependence of COPD exacerbation or mechanical ventilation to relative independence through participation in an inpatient pulmonary rehabilitation program (PRP). After receiving approval from the Institutional Review Board at Barlow Respiratory Hos-

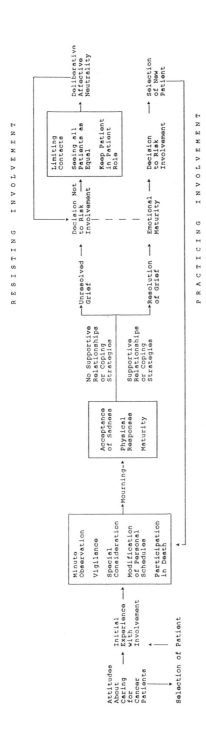

Figure 15.1. Conceptual Map of Phenomenon of Risking Involvement

pital, 15 patients who were participants in the PRP were interviewed in tape-recorded interviews either in the hospital prior to discharge or in their homes, about their experiences in the program. In addition, observations of the rehabilitation sessions were made several times each week for a period of 2 years. These interviews and observations were analyzed using constant comparative analysis to identify the core process in the experience of the patients.

The basic social process, "regaining control," was identified. It has four phases that parallel the constructs of situational sense of coherence: 1) *seeing a glimmer of hope,* in which the patient sees other patients making progress and thinks "Maybe I can do it also" (comprehensibility); 2) *getting with the program, in which the patient decides it is worth the effort (meaningfulness);* 3) *accepting mentoring,* in which the patient accepts the goals of the staff for self and submits to their direction (mutual plan of care); and 4) *taking charge,* in which the patient learns effective delegation and body listening to accomplish what is important to him or her (manageability). A conceptual map summarizes these phases (see Figure 15.2). Findings from this study provide information about patient responses to the rehabilitative process that will aid health care professionals in working collaboratively with them.

Strategies for Increasing Meaningfulness in Patients Experiencing Difficulties in Managing Illness Situations[5]

Barbara M. Artinian

Antonovsky (1987) defines meaningfulness as the "extent to which one feels that life makes sense emotionally, that at least some of the problems and demands posed by living are worth investing energy in, are worthy of commitment and engagement, are challenges that are 'welcome' rather than burdens that one would much rather do without" (p. 19). This construct has been modified by Artinian (1991) to describe the person's response to an actual illness experience, the situational sense of coherence (SSOC).

In nursing practice, the nurse often encounters a patient for whom the particular illness situation poses demands the patient is unable or unwilling to meet. It may be that the patient does not fully understand the situation, believes that resources are not available

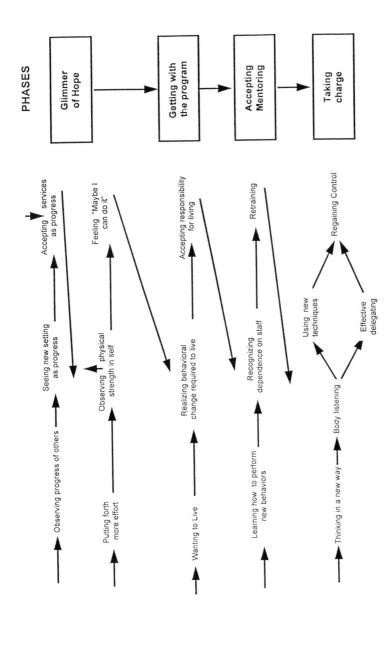

Figure 15.2. Regaining Control After Prolonged Mechanical Ventilation

263

for managing the situation, or has cultural beliefs that are in conflict with Western medicine. Any of these barriers can lead to an impasse. In some cases in which a lack of knowledge is the problem, the nurse could communicate information about the illness and community resources until the patient understands what could be done. However, a lack of knowledge may not be the problem. Even with an intellectual knowledge of the meaning of the situation, the patient may deny its seriousness, be convinced that there is nothing possible to be done to improve the situation, or because of cultural or religious values believe that nothing should be done or that the approach suggested by the health professional is not appropriate. This results in the need to negotiate values. It is impossible to progress to the next step, that of organizing behaviors to develop a mutual plan of care to meet the problem, until the patient and nurse agree on the goals of treatment.

This study is designed to learn about the mutual goal-setting process used by nurses and patients—to learn what the nurse does to negotiate with the patient to bring about attitudinal changes that may help the patient see the problem in a new light or assist the nurse to understand the problem in a new light from the patient's perspective. The focus is on attempting to discover what nurses do, where their knowledge comes from, and how they process it in negotiation with the patient. Chenitz (1984) categorizes this type of research activity as "surfacing nursing process." The research question is, "How do nurses interact with patients who do not accept the goals the nurse believes are beneficial for the patient?"

This will be an exploratory descriptive study using the constant comparative analysis methodology of grounded theory (Glaser, 1978). Nurses who use the Intersystem Model in their practice will be asked to describe interactions they have had with patients when mutual goal setting was problematic. Since nursing interventions take place in one-to-one interactions, the presence of a researcher during the interaction process would change the interaction. Therefore, the critical incident technique will be used to focus in-depth interviews about nurse-patient interactions, either in individual tape-recorded interviews or in focus groups. This technique is useful when trying to identify the content and characteristics surrounding interactional situations between individuals or groups of individuals. The word "critical" means that the incident must have had a discernible impact on some outcome: It must make either a positive or negative contribution to the accomplishment of some activity of interest (Polit & Hungler, 1983, p. 348). Participants will be recruited from announcements placed in the Azusa Pacific University annual alumni surveys or in the graduate student nurses' newsletter. The following format will be used for recording the critical incidents.

1. Select an incident in which your clinical judgment and the patient's expressed wishes led to different ideas about what should be done. How did you know there was a disagreement? What was the nature of the disagreement?

2. When you find that a patient does not accept the goals you believe are necessary for well-being, what indicators give you direction for further interaction? How would you be sure that you understand what problems the patient is expressing?

3. What additional information did you seek? What were your concerns at that time? Did this change your opinion about what should be done?

4. How did you go about attempting to secure the patient's cooperation? What negotiation attempts did you make? What were you feeling as you attempted to negotiate? What happened during the negotiation? What was the outcome?

5. How satisfied were you with the interaction? Is there anything you would do differently another time? Why?

Tapes of all interviews will be transcribed, and content analysis will be done. Categories will be developed and refined with the goal of capturing the "the data inherent in nurse-client interactions, which form the basis of nursing practice and theory development" (Chenitz, 1984, p. 206). The purpose of the analysis is to discover the basic social process underlying the experience of negotiating to develop mutual goals. When the process is identified, it will be discussed with nurses as a way to validate the theory and refine and expand it.

The Effect of Congruency of Expectations of Patient and Staff on Rehabilitation Outcomes of Patients With COPD[6]

Barbara M. Artinian

Chronic obstructive pulmonary disease (COPD) poses a major health problem for many elderly persons. Patients come to a pulmonary rehabilitation program (PRP) with different expectations for themselves. Success of a PRP for a particular patient depends on factors such as the extent of lung disease and knowledge about the illness, family support, and motivation, all of which are components of the situational sense of coherence (SSOC). SSOC is a measure of the integrative potential in the person's understanding of the illness situation, the way of looking at the situation, and the ability to marshal resources. It is hypothesized that individuals who have a higher SSOC will have more confidence in their ability to overcome a stressful situation and therefore greater expectations for success.

Success also depends on a knowledgeable staff that is able to accurately assess the rehabilitative potential of each patient. Therefore, patient outcomes may depend on the degree of congruence of patient and staff expectations.

This study, which combines quantitative and qualitative approaches, was developed to study the process of negotiation between patient and health care professionals to bring expectations into congruence. The study will be conducted at two outpatient pulmonary rehabilitation programs. Patients who have rehabilitative potential are admitted to the hospital-based outpatient pulmonary rehabilitation programs through referrals from their physicians.

The first purpose of this study is to compare the congruency of patient and staff perceptions of the patient's SSOC and expectations for achievement of goals at the time of admission to the program and at weekly intervals with patient outcomes at the time of discharge. A second purpose is to study the process of information communication, negotiation of values, and mutual goal setting that takes place between staff and patient to make expectations congruent. The following hypotheses will be tested.

1. Patients whose SSOC scores are more congruent with staff ratings will express goals that are more congruent with staff goals for them.
2. Patients whose expectations for achievement of goals are more congruent with staff expectations will be more likely to achieve the goals of the program.
3. Patients who achieve their expectations will express greater satisfaction with the pulmonary rehabilitation program.

A third purpose of the study is to develop a grounded theory related to patient-staff interaction associated with the process of increasing the SSOC of patients experiencing limitations imposed by COPD.

All patients who speak English who are admitted to the outpatient pulmonary rehabilitation programs during a 6-month period will be asked to participate.

Patients will score themselves on SSOC, dyspnea, and Pulmonary Functional Status level at admission and at weekly intervals, and on SSOC, dyspnea level, and satisfaction at discharge. They will be interviewed on admission and at weekly intervals to determine what goals they have for themselves and their expectations for achieving them. At discharge they will be interviewed about their perception of goal accomplishment. In addition, observations will be made by the principal investigators of patient-staff interactions during the rehabilitation sessions. Using the staff form of the SSOC instrument, the program director will score patients on SSOC on admission and discharge to reflect staff perception and will summarize the goals for the patient developed by the team at admission and at weekly intervals and evaluate their achievement at discharge. The associations between admission and discharge scores will be analyzed using *t* tests and ANOVA, and grounded theory methodology will be used to analyze qualitative data.

Permission has been obtained from the Institutional Review Board at Azusa Pacific University and the participating programs to conduct this study. It is expected that increased understanding of the mutual goal setting process described in the Intersystem Model should be of benefit to all COPD patients, their families, and the professionals who work with them.

Discussion

In the qualitative studies described previously, the major research focus is on how individuals and groups carry out their lives together and accomplish change rather than the what or why questions of quantitative research, although the descriptive and explanatory factors may be used as context. This asking of a different question, "What is the process that is taking place?" provides a holistic and process orientation to the research endeavor. This approach is in the tradition of Cooley, Thomas, and Mead, all of the early symbolic interactionist school, who considered "the subjective aspect of human behavior as a necessary part of the process of formation and dynamic maintenance of the social self and the social group" (Psathas, 1989, p. 5). The focus is on the holistic emergence of knowledge within the ongoing social process of the participants. Psathas comments that research that captures the reality of the participants can only be done when the researcher becomes an observer of the negotiations of everyday life. It is only in this way that we can understand how our subjects make sense of their lives in the real world.

Notes

1. Reprinted with permission from Milligan-Hecox (1996).
2. Reprinted with permission from Stilwell (1995).
3. Reprinted with permission from Artinian (1995b).
4. Reprinted with permission from Artinian (1995a).
5. Reprinted with permission from Artinian (1996b).
6. Reprinted with permission from Artinian (1996a).

References

Antonovsky, A. (1987). *Unraveling the mystery of health: How people manage stress and stay well.* San Francisco: Jossey-Bass.

Artinian, B. (1982). Conceptual mapping: The development of the strategy. *Western Journal of Nursing Research, 4,* 379-393.

Artinian, B. (1986). The research process in grounded theory. In W. C. Chenitz & J. Swanson (Eds.), *From practice to grounded theory: Qualitative research in nursing* (pp. 16-23). Menlo Park, CA: Addison-Wesley.

Artinian, B. (1991). The development of the Intersystem Model. *Journal of Advanced Nursing, 16,* 164-205.

Artinian, B. (1995a, April). *The process of regaining control: Outcome of a pulmonary rehabilitation program for patients with advanced COPD.* Paper presented at the annual meeting of the California Society of Pulmonary Rehabilitation, San Francisco.

Artinian, B. (1995b). Risking involvement with cancer patients. *Western Journal of Nursing Research, 17,* 292-304.

Artinian, B. (1996a). *The effect of congruency of expectations of patient and staff on rehabilitation outcomes of patients with COPD.* Unpublished research paper, Azusa Pacific University, Azusa, CA.

Artinian, B. (1996b). *Strategies for increasing meaningfulness in patients experiencing difficulties in managing stress situations.* Unpublished research paper, Azusa Pacific University, Azusa, CA.

Blumer, H. (1969). *Symbolic interactionism.* Englewood Cliffs, NJ: Prentice Hall.

Brammer, L. M. (1988). *The helping relationship.* Menlo Park, CA: Addison-Wesley.

Chenitz, W. C. (1984). Surfacing nursing process: A method for generating nursing theory from practice. *Journal of Advanced Nursing, 9,* 205-215.

Chenitz, W. C., & Swanson, J. (1986). *From practice to grounded theory: Qualitative research in nursing.* Menlo Park, CA: Addison-Wesley.

Cone, P. (1994). *A qualitative study of the experience of giving spiritual care.* Unpublished master's thesis, Azusa Pacific University, Azusa, CA.

Conger, M. (1991). *The efficacy of the nursing decision grid on delegation decision making skills of registered nurses.* Unpublished master's thesis, Azusa Pacific University, Azusa, CA.

Conger, M. M. (1994). The Nursing Assessment Decision Grid: Tool for delegation decision. *Journal of Continuing Education in Nursing, 25,* 21-27.

Glaser, B. (1978). *Theoretical sensitivity.* Mill Valley, CA: Sociology Press.

Glaser, B., & Strauss, A. (1969). *The discovery of grounded theory.* Chicago: Aldine.

Milligan-Hecox, J. (1996). *Understanding how patients with chronic obstructive lung disease manage their illness.* Unpublished research paper, Azusa Pacific University, Azusa, CA.

Polit, D., & Hungler, B. (1983). *Nursing research: Principles and methods.* Philadelphia: J. B. Lippincott.

Psathas, G. (1989). *Phenomenology and sociology: Theory and research.* Lanham, MD: University Press of America.

Ronai, C. C. (1992). The reflexive self through narrative: A night in the life an exotic dancer/researcher. In C. Ellis & M. Flaherty (Eds.), *Investigating subjectivity: Research on lived experience* (pp. 102-124). Newbury Park, CA: Sage.

Stilwell, K. (1995). *Communication with the ventilator-dependent patient.* Unpublished research paper, Azusa Pacific University, Azusa, CA.

Tourigny, S. (1994). Integrating ethics with symbolic interactionism: The case of oncology. *Qualitative Health Research, 4,* 163-185.

16

Connecting

The Experience of Giving Spiritual Care

Pamela Cone

Statement of the Problem
Significance
Conceptual Framework
Methodology
Basic Social Process: Connecting
Comparison of Psychosocial Care and Spiritual Care
Discussion
Conclusion

Spiritual care is an integral part of nursing that until recently has largely been ignored in research literature. Historically and conceptually, nursing is a caring profession that values wholeness and calls for care of the entire person. In practice, however, nurses have had a greater focus on the care of the physical rather than the spiritual being (Collition, 1981). Mickley, Soeken, and Belcher (1992), in their study on spiritual health, state that "although nurses have considered spiritual health as an important component of the total health of an individual, what it means to be spiritually healthy has not been conceptually clarified" (p. 267). This lack of clear definition of spiritual health results from a lack of commonality in the conceptualization of the spiritual aspect of the individual. According to Ruth Stoll (1989), there is universal concern about spirituality but "little universal consensus" (p. 5).

Statement of the Problem

Although there is current acknowledgment that the spiritual dimension in patient care should be addressed, nursing literature and research in the area of spiritual care are limited (Edgar, 1989; Shelly, 1993). The nursing diagnosis of spiritual distress has received mental assent from nursing professionals, but in actual practice little spiritual care is given. and the identification of spiritual distress in nursing care plans is rare indeed (Edgar, 1989; Shelly & Fish, 1988).

One major problem lies in the definition of the term *spiritual care*. Clarification of what constitutes spiritual care, in particular, the differentiation between the spiritual and the psychosocial, becomes increasingly important as nurses begin to address the need for spiritual care within the nursing profession. Unfortunately, the descriptions of this phenomenon are extremely diverse. This causes confusion among nurses, which in turn often leads to the misdiagnosing or overlooking of spiritual distress by the nursing profession. The purpose of this study was to explore the experience of providing spiritual care, with the goal of clarifying the difference between the giving of spiritual and psychosocial care.

Significance

When addressing an area in which there is confusion and even controversy, the significance of qualitative research lies in its ability to clarify the phenomenon under investigation from the viewpoint of those experiencing it (Chenitz & Swanson, 1986; Glaser & Strauss, 1967; Haase, Britt, Coward, Leidy, & Penn, 1992; Haberman-Little, 1991). This study sought to clarify the essence of spiritual care from a nursing perspective, using as the sample the faculty and students of a school of nursing. If this clarification will enable nurses to differentiate between spiritual and psychosocial care, this will be a significant contribution to the discipline of nursing.

Conceptual Framework

Spirituality is described in the nursing literature as the sense of self or personhood as well as that sense of relationship with a supreme being, however that term is defined by a particular creed or religion (Carson, 1989).

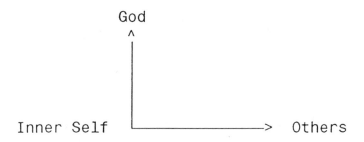

Figure 16.1. Vertical and Horizontal Relationships

Ellison (1983) writes that it is the integrative force between our "psyche and soma" (p. 332). Spirituality is not religiosity. It is an experiential reality rather than a system of ideas and principles. Religions tend to be culture bound, whereas spirituality is "transcultural, integral, and truly universal" (Rouch de Coppens, 1980, p. 18).

In the model of self used in this study (taken from the Intersystem Model developed at Azusa Pacific University School of Nursing), the person is represented by three concentric circles with the spirit at the center (Artinian, 1991). The spirit, which in its vertical dimension allows for the awareness of God, is the integrator of the person and moves freely across the psychosocial circle and the physiological or outer circle. In its horizontal aspect, it is directed toward others. According to Carson (1989), the "horizontal facet reflects and fleshes out the supreme value experiences of one's relationship with God through one's beliefs, values, life-style, quality of life, and interactions with self, others, and nature" (p. 7).

Each person, or system, has these three subsystems, with the vertical and horizontal dimensions emanating from the spiritual core. The interaction between systems, such as the nurse and client, is one focus of the Intersystem Model. This horizontal dimension is closely tied to the vertical. Although the interaction is of necessity expressed in many ways, including psychosocially and physiologically, actual spiritual care would be any care that enables patients to exercise their beliefs, values, practices, and affective responses and would be differentiated from psychosocial care by the enabling of the vertical facet of the person rather than the horizontal. It is the "supernatural aspect of spiritual care" that distinguishes it from other care (Karns, 1993, p. 5). Using the model shown in Figure 16.1, one can conceptualize more clearly the difference between the spiritual, or vertical, and the psychosocial, or horizontal, aspects of nursing care.

According to Phyllis Karns in her work "Building a Foundation for Spiritual Care" (1993), nursing has moved away from the understanding of the supernatural aspect of spiritual care as expressed by Florence Nightingale (1863). Nursing trends have moved toward the more humanistic views of spirituality as expressed by such theorists as Sister Calista Roy and Jean Watson, who include the spiritual in "all that transcends the material" (Karns, 1993, p. 3).

The Christian perspective of spirituality, in which the spiritual dimension is central and integrative and focuses on one's relationship with God through Jesus Christ, is espoused by the Azusa Pacific University School of Nursing. It is from this perspective that the Intersystem Model has been developed (Artinian, 1991). In this model, spirituality, or the spiritual subsystem, is seen as central to the person. The spiritual subsystem contains the beliefs, values, practices, and affective responses that govern the person's interactions.

Spiritual care does not require a shared religious or philosophical worldview (Carson, 1993). What is necessary is an empathetic concern and willingness to accept the other person's views and to assist that person along his or her own spiritual journey (Edgar, 1989). The nurse's own level of comfort with spiritual issues does affect the ability to personally render spiritual care (Dickinson, 1975; Edgar, 1989); actually acknowledging and identifying the area of spiritual distress, however, is a major step in the giving of spiritual care. Research on the nurse's role in providing spiritual care is limited. To find common ground, experiential studies of the concept must be undertaken.

Methodology

This was a qualitative research study using the constant comparative analysis of the grounded theory method (Chenitz & Swanson, 1986). This method of field research is used to identify core categories and the basic social processes (BSPs) that are being experienced by the individuals in the situation under study (Glaser, 1978). According to qualitative researcher Holly Skodol Wilson (1991), this type of nursing research "allows us to weave a tapestry from the threads that make up the dailiness of our practice lives" (p. 281).

Design

Using qualitative research methodology within the symbolic interactional perspective (Blumer, 1969), the researcher chose an exploratory design to determine what spiritual care and the experience of giving spiritual care mean to those administering it. In addition to studying questionnaire responses and spiritual care episodes submitted by the sample population, the researcher conducted dialogues on the subject with focus groups and drew examples from firsthand reports of spiritual care found in the literature. Data from all these sources were coded when collected to develop categories that provided direction for further data collection. Theoretical memos were written that aided in the discovery of the basic social process.

Sample

The faculty and student body of the school of nursing of a private Christian university were the sample for this study. The setting was Azusa Pacific University School of Nursing (APU-SON) located in Azusa, California, a part of greater Los Angeles. Using systems theory as a frame of reference, APU-SON can be described as a system with four subsystems: faculty, graduate students, undergraduate students, and staff. The first three subsystems were included in the target sample population because of their nurse-patient experience. Research packets went to 35 faculty members, 100 graduate students, and 200 undergraduate students, making a total of 335. Group dialogues were conducted in 15 classes covering all levels of study. Six classes were at the graduate level, 2 of which were in the first year of study. Within the 9 undergraduate classes, 1 class of 30 students was at the first level of nursing education, 2 classes of approximately 20 each were at the second level, the third level of study included 2 classes with about 20 and 1 with 10 students, and the fourth level had 2 classes of 10 and 1 of 20 students.

The criteria for inclusion in this study were language, age, and informed consent. The subjects were English-speaking faculty and students 18 years or older who agreed in writing to be in the study. A written explanation of the study and its purpose were included on the consent form. The various aspects of the study, such as the "fleshing out" of the spiritual care episodes and the audiotaping of group dialogues, were discussed so that informed consent could be ensured.

Confidentiality was preserved by assigning a number to each respondent and using that number to identify the sources of information during data analysis. No names were represented anywhere other than on the consent forms, which were filed separately from the data to provide the respondents with anonymity. In addition, all taped group dialogues were transcribed by the researcher.

Data Collection

The data collection process of this study used a three-pronged approach. First, a questionnaire was developed requesting individual definitions of what actually constitutes spiritual care as well as how it differs from psychosocial care. Second, guidelines for descriptions of personal interactions identified as spiritual care episodes were specified. These two forms, along with a letter of explanation, a consent form, and a demographic data form, made up the research packet that was given to the faculty, graduate students, and undergraduate students of APU-SON. Responses to the research packets were returned to the APU-SON office.

The research packet had an overall response rate of 18% with the faculty returning 22%, the graduate students 20%, and undergraduates 14%. The 6% male responses reflected a slightly higher percentage than the actual percentage of male students (5%) at APU-SON. The demographic categories of ethnicity, marital status, and age reflected extreme diversity, but no correlation was found between those categories and the concept under study.

Finally, the third approach involved the researcher taking the question into 15 classrooms for spiritual care dialogues. Three questions were posed to the students: (a) What is spiritual care? (b) In what way, if any, is spiritual care different from psychosocial care? and (c) Have you ever given spiritual care to a patient? Those who answered yes to the third question were asked to describe the interaction. These dialogues, which took place in the various classes at all levels of the program, were taped and later transcribed for analysis. Six graduate classes and nine undergraduate classes participated in the study, with faculty occasionally joining in the discussion.

Data Analysis

Constant comparative analysis of the data was begun as soon as the first set of responses was received. On the basis of the questions posed, the written responses, and group discussions concerning the topic of spiritual care, responses were divided into two primary categories for analysis. The first was the actual definition of spiritual care, and the second was the possible differentiation between psychosocial and spiritual care. Categories emerged within each topic as data were compared. Memoing was then used to describe the emergent categories. Data were analyzed and compared until these categories were saturated. Single, extremely different responses were set aside when comparisons showed no supportive data; for example, only one person said that there is no difference between psychosocial and spiritual care, whereas all the other responses described the difference in various ways. Verification of categories occurred as patterns reappeared among the data and a basic social process, connecting, was identified. Categories were then readjusted to fit the emerging theory. It is noteworthy that the spiritual system categories of the Intersystem Model—that is, beliefs, values, practices, and affective responses—were validated by the data processed regarding spirituality. This is discussed under Assessing Spirituality (p. 278).

Basic Social Process: Connecting

The basic social process of connecting has three stages (see Figure 16.2). The first stage, accepting, includes assessing spirituality, establishing trust,

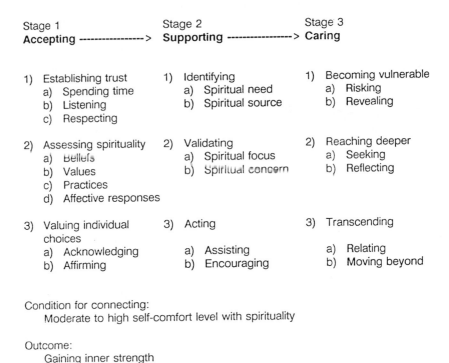

CONNECTING

Contexts:
    Sense of the supreme
    Inner self
    Situational specificity

Stage 1 **Accepting** ---------------->	Stage 2 **Supporting** ---------------->	Stage 3 **Caring**

1) Establishing trust
    a) Spending time
    b) Listening
    c) Respecting
	1) Identifying
	    a) Spiritual need
	    b) Spiritual source
		1) Becoming vulnerable
		    a) Risking
		    b) Revealing

2) Assessing spirituality
    a) Beliefs
    b) Values
    c) Practices
    d) Affective responses
	2) Validating
	    a) Spiritual focus
	    b) Spiritual concern
		2) Reaching deeper
		    a) Seeking
		    b) Reflecting

3) Valuing individual
    choices
    a) Acknowledging
    b) Affirming
	3) Acting
	
	    a) Assisting
	    b) Encouraging
		3) Transcending
		
		    a) Relating
		    b) Moving beyond

Condition for connecting:
    Moderate to high self-comfort level with spirituality

Outcome:
    Gaining inner strength
    Being uplifted in spirit

Figure 16.2. The Process of Connecting

and acknowledging individuality. The second stage, supporting, involves identifying, validating, and acting. The third stage, caring, involves becoming vulnerable, reaching deeper, and transcending. A condition for moving into this final stage was a moderate to high level of comfort with giving spiritual care based on the nurse's previous self-exploration in the area of spirituality.

The basic social process of connecting occurred within three contexts. The first was a sense of the supreme, or that which transcends. A few instances were given in which the patient's sense of the supreme was within and was expressed as a sense of control over personal destiny. Most, however, described the sense of the supreme as a "higher power," although many used the term "God." This transcendent quality was considered to be the major element identifying nursing care as particularly spiritual in nature. One respondent stated, "To me, spiritual care is finding out from them where they derive their inner strength, from their inner spirit, from a higher power, or whatever."

The second context was the inner self, inner spirit, or spiritual core, also referred to as the "soul." This was described as the "deepest, innermost part of the person." According to the data, the inner self of both the patient and the nurse is a part of spiritual care. Both the nurse and the patient must be willing to open up those things that are most private and deep for spiritual care to be given. "It's got to be mutual," said one respondent.

The third context was situational specificity. Discussions of deep inner thoughts and feelings do not easily take place with bystanders as an audience; therefore, spiritual care usually occurred when the nurse and patient could interact one on one, although on rare occasions another person was present who was identified as a part of the situation. One nurse said, "I knew I had to get him off alone so that we could talk about it." Disclosure of spiritual needs came within the context of issues, such as life and death, health and illness, culture and ethnicity, and meaning and purpose, that were specific to the situation. Spiritual care occurred at many times, places, and levels, and each situation was unique and meaningful in its own way. This was described by one nurse in the following way:

> Sometimes it's about the meaning of life, because in those situations their meaning of life really changes [about] what they're going to do with their lives and all about their, not just psychological well-being, but really that very essence of meaning—of where they belong in the world.

The outcome of the spiritual care interaction was seen as the uplifting of the patient's spirit and the gaining or enhancement of inner strength.

Stage 1: Accepting

The first stage, accepting, was described in the data as learning about and understanding the patient's spiritual frame of reference and upholding that person's right to act on personal beliefs and values. This stage includes three phases: establishing trust, assessing spirituality, and valuing individual choices.

Establishing Trust

Establishment of trust is important to the nurse regardless of whether the nature of nursing care is physiological, psychosocial, or spiritual. There is a certain element of trust in the profession itself, but individual trusting relationships must be established for issues of deep concern to be discussed.

According to the respondents, the nurse must first spend time with the patient, get to know the patient as a person, and indicate an interest in that person's concerns. Also involved is listening, a part of therapeutic communication, because as one nurse said, "you have to listen to let them know you think what they're saying is important." Spending time and listening were described as part of the process of establishing trust as well as respecting the patient's person, beliefs, choices, and rights. One student nurse mentioned how important it is not to "push" but to go at the patient's pace. Another said, "It's a matter of respecting."

Assessing Spirituality

Assessing spirituality, the second phase of accepting, entails the exploration of the patient's inner self and included the areas of beliefs, values, practices, and affective responses. Personal beliefs about God, life, death, and the hereafter guide feelings and actions concerning events and crises. While relating several experiences of medical intervention with patients adhering to a particular creed, one respondent explained:

> If it was on an emergency basis, and they had to have one [blood transfusion], and there was no one there to not do it, to tell [the staff] they couldn't do it, um, we'd want to counsel them afterwards because they would feel that they have done something against their religious beliefs . . . [even though] they didn't really have a choice if they were unconscious or something, you know. They would want to have some counseling after, because that can cause a lot of spiritual distress. It seems so minor to us, but to them it's very important.

Values of truth, love, meaning, and hope, among others, also guide the person's life. These values affect how the patient views crisis and change and the effect they have on the person's life and situation. For example, a student reported that one patient described her hospitalization as a time of quiet resting in God's love. Another student reported that the patient asked how a loving God could allow her to experience such pain.

Religious practices are another part of individual spirituality. Although they are not the essence of spirituality, they are the working out of personal beliefs and values and therefore have a direct impact on a patient's actions and interactions. One nurse discussing a spiritual care episode in which her actions were directed by a patient's belief system explained, "because they are very precise in their beliefs, like they cannot do anything on the Sabbath, . . . there's a [particular] way to do it."

The final aspect of assessing spirituality, affective responses, is the area that deals with the outward expression of the inner spiritual experience. When a patient's situation comes in conflict with beliefs, values, and practices, the affective response is directly related to spiritual issues rather than psychosocial ones. Tears of grief on the loss of childbearing potential, explained one nurse discussing a spiritual care episode, were a result of the patient asking the question, "Why did God let this happen to me?" The patient's belief concerning God's love for her was challenged by loss. Another respondent reported that one patient's peaceful countenance when faced with a critical surgery was due to a strong belief in God's continual presence as evidenced by the statement, "The Lord is with me. He's always with me!"

Valuing Individual Choices

The third phase of accepting is valuing. Valuing individual choices was described as communicating to the patient the value of the patient's personal spirituality as well as the total acceptance on the part of the nurse of the patient's spiritually driven choices.

The first aspect of this was seen as acknowledging the patient's belief system in a way that was not judgmental. This was particularly stressed when the nurse and patient belief systems were completely different. One discussion about the hospitalized patient with particular beliefs and practices centered on the patient's concerns about being judged or ignored by the staff. The patient stated,

> Sometimes I feel like my beliefs are being ignored. You all put up all kinds of Christian holiday decorations and you pay no attention to the fact that I'm Jewish and I don't believe all that stuff! You think I'm all wrong, so you just let the Sabbath go by without a word about how I feel.

Affirming the acknowledged belief system and resulting choices was the second aspect of valuing. A nurse's efforts to discover what lay behind choices was considered an important way of affirming the value of individual spirituality. One nurse, relating an experience with a hospitalized patient who did not want a particular room, stated,

> His daughter was telling me that that's a very unlucky number in [their country]. She says—I know it's superstition. She was trying to play it off as unimportant, but it wasn't unimportant to him or to them. And so I said, no, if it's important to you, it's important to me. We'll change the room. Again it's letting him and others know that their beliefs are important.

Acceptance was the major focus of several group dialogues. It was stated many times that the nurse must have total acceptance of the patient's individual spirituality for the patient to be willing to share deeper concerns:

Being accepted by the staff and being accepted by people [is very important] and part of all religious experience is being accepted by God. If we don't feel that we have that, our relationship with God will be severely compromised.

The discussion of acceptance usually brought up the category of spiritual support.

Stage 2: Supporting

The second stage of the process, supporting, was discussed as being one of the nurse's major roles. Support was described as an interrelatedness between the nurse and the patient that grew out of acceptance. It was viewed as an empathetic concern for the patient's welfare. Spiritual support was often at first identified as spiritual care. Through group discussion, the conclusion was reached that it is a part of the process and not the process itself. This action could only take place if the patient was willing to allow the nurse to move through the phases of the process. Supporting was seen as having three phases: identifying, validating, and acting.

Identifying

Identifying, the first phase of supporting, involves the identification of the spiritual need as well as the spiritual source of strength. Observation as well as listening were discussed as ways of discovering the spiritual need of a patient. In addition, a general knowledge of cultural mores, ethnic concerns, and religious differences often assisted the nurse in discovering a spiritual need. As one nurse stated, "it might be religion in some, or it might be . . . [a cultural group] that had a completely different belief system."

It was interesting to note that the dialogue participants placed a greater emphasis on the second aspect of identifying, that of identification of the source of inner strength, than on the actual need itself. This was mentioned over and over again as an important part of giving spiritual support. One nurse discussed finding out "where they derive their inner strength . . . and is that inside thing happening." Another stated, "You need to help them identify, find out, from where they derive their strength—from their inner self, from a higher power, or whatever—and whether that's satisfactory to them."

Validating

Once identification of the need and source of inner strength had occurred, the second phase, validating, began. The spiritual focus of the patient was explored to discover if the source was meeting the need. "It's having a patient's heart be right and their mind be at peace to rest and to heal," said one nurse.

The nurse then validated the spiritual concerns that had been raised, assuring the patient of the importance of every concern whether or not the need was being met. One nurse, discussing the value of this process with a depressed patient, expressed the idea that this level of care could explore

> the thoughts that have evoked these emotions in an individual, and then they have to reckon with them. They may see that their supports are diminishing, and these could even be in terms of what they consider to be their God or their driving force.

Acting

Acting, the third phase of supporting, was one area that most nurses found very satisfying. Two categories of action emerged from the data. These were assisting and encouraging. Assisting was described as actions such as praying, reading scripture, calling clergy, and adjusting dietary regimes to fit religious practices. Many such experiences were shared: "I ask, can I pray with you," "I can share my little devotional book—it's neat," or "Sometimes I've had patients who'll have a Bible sitting next to their bed, and I'll just ask them if they want me to read to them." This was perhaps the most frequently mentioned and commonly practiced type of spiritual intervention identified in the data.

Encouraging was also mentioned frequently. One aspect of encouraging was providing hope. This was seen as directing the patient toward that which provides meaning and purpose in the midst of difficult circumstances. Hope was used in the deep sense of good eventually resulting rather than just the sense of wishful thinking. Sometimes the more tangible act of bringing good news provided hope, and sometimes the hope came from faith and trust. The importance of this was explained in the statement that "fundamental to all of us is wanting peace, wanting hope."

The other aspect of encouragement is providing comfort. Comfort was raised as a psychosocial issue, but most participants agreed that there is a comfort that goes beyond the touch of a hand and the speaking of a sympathetic word. The words "console" and "comfort" were brought up many times in relation to difficulties that had no immediate solution. Some stated that just being with the patient at such a time could provide comfort, but most felt that spiritual comfort included the vertical dimension of a sense of the supreme. One student nurse stated that many of her patients "just take comfort in their relationship with God . . . if they're distressed or something like that . . . so I believe a lot of times they already know and you just have to tap into that belief."

Many nurses stated that spiritual support was as far as they usually go with their patients. Moving beyond this required evaluation of oneself and a moderate to high level of self-comfort with spirituality and spiritual issues. Many expressed a certain fear of entering a deeper level of spiritual care: "I'm afraid I won't know what to say," or "What if they ask me something I can't answer?" Others expressed a fear of being rejected and a desire to "not overstep a boundary." The general consensus, however, was that spiritual support, although extremely important, was not the end itself but was a part of the process in providing spiritual care.

Stage 3: Caring

True spiritual care was identified by most participants in this study as a caring that goes beyond and transcends the normal nurse-patient interaction. A special "interconnectedness" occurred in this process. Caring involves becoming vulnerable, reaching deeper, and transcending.

Becoming Vulnerable

Perhaps the most difficult phase of the process of caring is that of becoming vulnerable. According to the data, there is an openness required on the part of the nurse, a moving out of one's "comfort zone." This phase had two aspects: risking and revealing.

Risking is the inner step the nurse has to take, whereas revealing involves a willingness to share deep thoughts and feelings with the patient. It is a "willingness to enter into the intimacy of a relationship where those issues can be discussed," said one. Some nurses felt uncomfortable with the idea of revealing and sharing common ground with the patient, and others stated that it added a special touch, a "warm closeness" to the relationship. Several felt that it was necessary to the process of connecting. One student nurse said she often made a kind of connection, explaining,

> If my patient had some Bible verses picked out, then I just told them that I was a Christian, too, and that just seemed to make a difference to them. . . . It just seemed that knowing that I was a Christian meant a lot.

Reaching Deeper

Once the nurse revealed an openness and vulnerability to the patient, it was then possible to reach deeper into the patient's inner self, seeking the real spiritual need. The inner self was described as the deepest, innermost part of the person. The terms *inner spirit, spiritual core,* and *soul* were also used in the data. One nurse stated, "all it takes is just such a gentle probing and the whole thing is opened up, and then you can start getting them working on it." In discussing her interaction with a patient who had described a spiritual experience of a loving presence in her hospital room, one nurse said, "I felt like an honored receptacle, allowed to hold the information, but not expected to do anything about it." No more was needed at that time.

Reflecting is a natural part of therapeutic communication and, according to the study participants, encouraged the patient to search his or her own belief system for answers. It also enabled the nurse to help the patient in "processing information and emotions and feelings and reactions" so that clarification of the deep spiritual needs was possible. One nurse said, "I was glad she wanted to use me, and that she seemed comfortable exploring her questions with me."

Transcending

Caring not only involves becoming vulnerable and reaching deeper but also involves transcending. This phase includes relating—that is, being with and "sharing burdens" with the patient—and also moving beyond. Being with the person in his or her hour of need was described as more than physical proximity or even psychosocial interaction; is being spiritually attuned so that there is mutual focus on the inner work. It involves a connection between the deep inner being of the nurse and the patient that is beyond normal communication and sharing.

Moving beyond the normal barriers and ministering to the human spirit was viewed as the final aspect of this phase. "It's being in tune to the very core, the very existence, the very nature or personhood of that person," explained one nurse. "It transcends everything that's natural and is really supernatural," stated another. One nurse, describing a situation in which a fellow staff member could not let go of his distress over a difficult patient death, related her spiritual intervention with him at this level, ending with, "I felt that this situation was a deeply spiritual experience for both of us. It went beyond psychosocial care and touched on the uncomfortable area of accountability and forgiveness that we tend to shy away from."

This basic social process, connecting, had an outcome of uplifting the spirit of the patient or enabling the patient to gain "compelling inner strength," or both. Although spiritual acceptance is necessary and spiritual support is vital, they are only a part of this process. Spiritual caring moves the nurse-patient relationship into something beautiful and rewarding to both. One nurse related her experience:

> This mother and I had really connected. We had such lovely interactions together throughout the entire day sharing the goodness and strength of the Lord. I really felt I had allowed her an avenue to open up her hurts and be honest as well as pull from the inner strength she possessed in the Lord.

ENTWINED:

spiritual care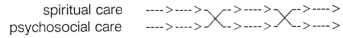
psychosocial care

CONTINUUM: ----> psychosocial ---->----|----> spiritual ---->

INTERSECT: spiritual care

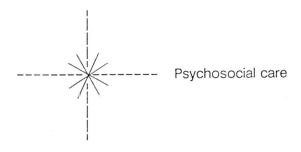

Psychosocial care

Figure 16.3. Typologies of Nurse-Patient Interaction

Comparison of Psychosocial Care and Spiritual Care

The other major topic of discussion in this study was the difference between psychosocial and spiritual care. Only one person wrote that, because we are spiritual beings, spiritual care encompasses all care that nurses provide. Everyone else expressed some sort of difference between psychosocial and spiritual care; they simply disagreed on the description of that difference.

Although the participants in this study differed as to how the connecting should be described in terms of spiritual and psychosocial care, there was general agreement that something special occurred when the nurse and patient made that spiritual connection. Most said that there was a point of "conscious decision" to move beyond into that deep inner area. One described an "internal tremor that continues until the interaction is over," explaining that it is "an internal indicator of how important whatever is going on really is."

Three major categories concerning the relationship of psychosocial and spiritual care emerged from the data. These categories, although not parallel, describe the views expressed in the written and taped data. The following types of relationships were described: entwined, a continuum, and intersect (see Figure 16.3).

Although the majority of the data supported the idea of a clear difference between these two types of care, some 14% of the verbal and 10% of the written responses indicated that the care is so interconnected that it is impossible to separate them. Some said they go "hand in hand," whereas others described the entwined nature of vines as a picture of this concept.

A significantly large number of respondents—29% of the verbal and 40% of the written—felt that these two types of care are on a continuum. The interaction moves along the psychosocial aspect of the continuum toward the spiritual. Some felt that the shift into spiritual care is somewhat nebulous and difficult to pin down. One said that the "interaction flows back and forth between the two types of care" so that it is impossible to tell when the nurse actually moves from one to the other. Others said that there is an "overlap," whereas the majority expressed the idea of a clear point of change. Most said that all at once they "knew" that they had moved into a deeper realm. Intuitivity was discussed along with this concept. That "deep inner knowledge that something is so" is a part of the giving of spiritual care.

The largest number of written (50%) and verbal (57%) responses expressed a three-dimensional view of intersecting concepts. The idea was expressed that the physiological, psychosocial, and spiritual aspects of the person are levels that go deeper and deeper into the inner being. The interaction between the nurse and patient causes an intersecting to occur between the subsystems of the nurse and the patient. When the intersection occurs at the spiritual core, the nurse becomes sensitive to or "in tune" with that person's spirit, with the "very nature or personhood of that person." Again, there is a point at which this occurs, and the nurse intuitively knows when spirits have touched. In this conceptualization of spiritual care, there are levels of even the actual giving of spiritual care. Supporting the person's beliefs and practices by reading scripture or praying can be at a more superficial level or at a very deep level, depending on the point of intersection or "connection." One nurse shared this experience with a father whose son was in critical condition:

> I pulled him aside, and I said let's go outside. Let's walk around and let's talk. . . . It's okay to tell me what's going on in you. At this point I want you to help me understand what's going on in your mind . . . and he just began to massively unload, unload, unload. And I sat there way into the hours of the night holding his hand waiting until we heard what was going on with his son in the OR. And at this point I connected with him. I sensed an incredible plea for spiritual support. It was beyond the coffee and the walking and the hand holding. It was connecting with him as a father, as a person, and where he was in relationship to mortality and I knew I was going into that realm because I was exhausted. It was emotionally draining. But I sensed, I knew, and he sensed interconnectedness.

Discussion

Data analysis revealed several strengths and weaknesses in the data collection process. The three-pronged approach was good because it provided a basis for comparative analysis between the groups of data, thereby enabling the basic social process to emerge from the data. The questionnaire provided data concerning the concept of spiritual care, whereas the spiritual care episodes form provided examples of providing spiritual care for comparative analysis. The dialogues provided background information for both of those aspects for study and had the advantage of being both heard and read by the researcher. This enabled the researcher to catch nonverbal cues and nuances of speech, adding richness to the data. One example of this was the noting of tears in one nurse's eyes as she shared her episode. Another advantage noted with the dialogues was the ability to seek immediate clarification of the intent of the speaker during discussions of the concept of spiritual care.

The overall response rate of 18% signaled a weakness in the timing of the data collection process: The research packets were distributed shortly before the end of the semester. Another weakness was the failure to send out reminder notices to the recipients of the packets. Also, an interest level bias might be detected if one considers that those with a higher interest-level in the subject may have made a greater effort to overcome time constraints and participate in the study.

Other strengths and weaknesses of the study were brought to light as responses were compared and analyzed. Many students asked if the findings would be reported because this was an area of high interest to them. A supportive faculty member expressed a desire to incorporate relevant data in nursing education. On the other hand, the researcher was told several times that the questions were difficult to answer, possibly because of the nature of the study, which touched deep concerns and was therefore an uncomfortable subject for some. In addition, the guidelines for writing the spiritual care episodes were very detailed and were described by some students as "rather intimidating." Nevertheless, of the 60 responses, 26 included descriptions of spiritual episodes. Only 2 respondents reported a low level of self-comfort when giving spiritual care; most respondents reported that they were moderately comfortable, and 13 were highly comfortable.

One wording problem concerning academic status was pointed out by some undergraduates who are first-level nursing students but second- and third-year college students. Another wording problem lay in the religious preference category. In an institution that encourages religious freedom within the Christian faith, it would have been preferable to have a category labeled "Christian" rather than the traditional separation of Catholic and Protestant.

Several responses used the "Other" category to denote a nondenominational Christian.

It was interesting to note that the demographic data did not support a positive correlation between the level of self-comfort when giving spiritual care and a stated religious preference. Instead, the data from the spiritual care dialogues indicated that the relationship lies between self-comfort with personal spirituality and the ability to comfortably provide spiritual care. Those who stated that they have explored their own sense of spirituality and come to a place of inner peace all voiced a high level of comfort with spiritual care.

The term *connect* was mentioned many times in both the written and the taped data. There was a sense of transcendence brought up with this concept. In other words, this connection is between the deep inner being of the nurse and the patient in a way that is beyond normal communication and sharing. It is the key to providing spiritual care.

Conclusion

Clearly there is a wide variety of views concerning spiritual care even within this small sample population in which spiritual issues are freely and frequently discussed. When all the threads of thought are drawn together, however, the basic social process emerges as connecting. The interaction between the nurse and patient becomes spiritual care when the nurse connects on a deep, transcendent level with the patient in such a way as to provide support and enablement for the patient's beliefs, values, practices, and affective responses with the outcome of uplifting the spirit and enhancing the inner strength of the patient in that particular situation. Further research on this important topic is recommended. This study should be replicated in other settings and with other sample populations to validate connecting as a substantive theory of spiritual care.

References

Artinian, B. (1991). The development of the Intersystem Model. *Journal of Advanced Nursing, 16,* 194-205.

Blumer, H. (1969). *Symbolic interaction: Perspective and method.* Englewood Cliffs, NJ: Prentice Hall.

Carson, V. B. (1989). *Spiritual dimensions of nursing practice.* Philadelphia: W. B. Saunders.

Chenitz, C., & Swanson, J. (1986). *From practice to grounded theory.* Menlo Park, CA: Addison-Wesley.

Collition, M. A. (1981). The spiritual dimension of nursing. In I. L. Beland & J. Y. Passos (Eds.), *Clinical nursing* (pp. 492-501). New York: Macmillan.

Dickinson, C. (1975). The search for spiritual meaning. *American Journal of Nursing, 75,* 1789-1794.

Edgar, M. A. (1989). *An investigation of the relationship between nurses' spiritual well-being and attitudes of their role in providing spiritual care.* Unpublished master's thesis, California State University, Department of Nursing, Los Angeles.

Ellison, C. W. (1983). Spiritual well-being: Conceptualization and measurement. *Journal of Psychology and Theology, 11,* 330-340.

Glaser, B. (1978). *Theoretical sensitivity.* Mill Valley, CA: Sociology Press.

Glaser, B., & Strauss, A. (1967). *The discovery of grounded theory.* Chicago: Aldine.

Haase, J. E., Britt, T., Coward, D. D., Leidy, N. K., & Penn, P. E. (1992). Simultaneous concept analysis of spiritual perspective, hope, acceptance and self-transcendence. *IMAGE: Journal of Nursing Scholarship, 24,* 141-147.

Haberman-Little, B. (1991). Qualitative research methodologies: An overview. *Journal of Neuroscience Nursing, 23,* 188-190.

Karns, P. S. (1993). Building a foundation for spiritual care. In J. A. Shelly (Ed.), *Teaching spiritual care.* Madison, WI: Nurses Christian Fellowship.

Mickley, J. R., Soeken, K., & Belcher, A. (1992). Spiritual well-being, religiousness and hope among women with breast cancer. *IMAGE: Journal of Nursing Scholarship, 24,* 267-272.

Nightingale, F. (1863). *Notes on nursing.* London: Spottiswoode.

Rouche de Coppens, P. (1980) *The spiritual perspective: Key issues and themes interpreted from the standpoint of spiritual consciousness.* Washington, DC: University Press of America,

Shelly, J. A. (Ed.). (1993). *Teaching spiritual care* (2nd ed.). Madison, WI: Nurses Christian Fellowship.

Shelly, J. A., & Fish, S. (1988). *Spiritual care: The nurse's role.* Downer's Grove, IL: InterVarsity Press.

Stoll, R. (1989). The essence of spirituality. In V. B. Carson (Ed.), *Spiritual dimensions of nursing practice.* Philadelphia: W. B. Saunders.

Wilson, H. S. (1991). Identifying problems for clinical research to create a nursing tapestry. *Nursing Outlook, 39,* 280-282.

Name Index

Albrecht, M., 175, 176
Allensworth, O., 133
Andrews, H. A., 9, 99
Angel, R., 88
Antonovsky, A., 3, 9, 18, 19, 20, 23, 24, 31,
    38, 39, 131, 199, 246, 262
Apter, M. J., 85, 87
Archer, J., 219
Arnold, M. B., 86
Artinian, B., 10, 15, 23, 32, 156, 178,
    195, 228, 246, 248, 257, 260-267,
    272
Auday, B. C., 166

Baker, L., 139
Bandura, A., 217
Bartter, F. C., 68
Becker, M. H., 39
Belcher, A., 270
Bem, S., 133
Benner, P., 47, 227, 252
Bennoliel, J. Q., 98
Bergman, P. G., 75
Berliner, R. W., 75
Bernard, C., 72
Bertalanffy, L. von, 8, 64
Bevis, E., 204, 210-212
Birch, H. G., 90
Blanchard, K. H., 159
Blum, M., 133
Blumer, H., 33, 49, 50, 255, 258, 272
Bond, M., 237-239

Bonham, L. A., 222
Boss, P., 47, 144
Bowers, I., 54
Bracht, N., 175
Brammer, L. M., 260
Brandon, K., 167
Braun-Menendez, E., 75
Brettle, R. P., 70
Brewer, M., 133
Britt, T., 271
Brookfield, S., 221, 222, 223
Brown, L. N., 228
Brown, M. A., 98
Brown, S. J., 2, 194, 197
Brown-Sequard, C. E., 72
Budd, K. W., 45
Burck, J., 108, 114
Burnard, P., 53
Burr, W., 24, 131
Buss, A. J., 89
Byrne, J., 133

Cantril, H., 88
Caplan, C., 55
Carson, V. B., 114, 271, 272
Carter, E., 135
Casey, K. L., 66
Chafetz, M. D., 73
Chalmers, K., 43
Chavis, D. M., 93
Chenitz, C., 271, 272
Chenitz, W. C., 256, 257, 264, 265

289

Subject Index

About the Authors

Barbara M. Artinian, RN, PhD, is Professor in the School of Nursing, Azusa Pacific University, Azusa, California. She teaches courses in community health nursing, family theory, nursing theory, and qualitative research methodology. She has written a nursing model, the Intersystem Model, which is the model used for the undergraduate curriculum at Azusa Pacific University. Her research studies include goal setting in a respiratory rehabilitation program, the experience of living with chronic obstructive pulmonary disease, and the giving and receiving of spiritual care.

Pamela Cone, RN, BSN, MSN, is Adjunct Professor at Azusa Pacific University School of Nursing, where she teaches stress theory and fundamentals of nursing. In addition to being a member of Sigma Theta Tau, she is currently a member of the board of trustees of Université Chretienne du Nord Haiti and an honorary member of the Glendora City Council. Her research in the area of spiritual care has become a nationwide study.

Margaret M. Conger, RN, MSN, EdD, is Associate Professor at Northern Arizona University. She is former Director of Education at Humana Hospital, Huntington Beach, California. While in that position, she was instrumental in developing educational needs assessments and programs for preparing RNs to work in partnership relationships with licensed practical nurses and nursing assistants. She has a strong background in the biological sciences and incorporates that understanding into her current interests in clinical decision making and critical thinking.

Ilene M. Decker, RN, MS, is Assistant Professor of Nursing at Northern Arizona University. She has extensive experience in nursing as a practitioner and an administrator in the areas of critical care, emergency, and burn therapy

303

nursing. She is currently pursuing doctoral studies in nursing at the University of Arizona. Her interests include instrument development and philosophy. Her current research interests are in critical thinking of nursing students and moral development of older adults.

Sandra G. Elkins, RN, EdD, is Assistant Professor of Nursing at Northern Arizona University. She has extensive experience in nursing as a clinical specialist in craniofacial and reconstructive plastic surgical nursing and adult nursing, as staff development educator and department director, and in teaching at the diploma and baccalaureate levels. Her current area of research is the ethical development of professional nurses using a grounded theory approach.

Julia D. Emblen, RN, PhD, is the nursing department chair at Trinity Western University, Langley, British Columbia, Canada. Her research interests include professional values, suffering—different religious views, and spiritual distress—developing assessment instruction

Margaret England, RN, PhD, is Associate Professor and Chair of the Division of Nursing at Keuka College, Keuka Park, New York. Her recent research has focused on the modulation of caregiver strain of adult offspring of Appalachian descent and the development of a schema for coding therapeutic conversation.

Phyllis Esslinger, RN, MSN, is Director of the Graduate Nursing Program and Associate Professor of Nursing at Azusa Pacific University. Her areas of expertise include pediatric nursing, community health nursing, research methodology, and nursing theory. Her research interests include measurements of the situational sense of coherence and validity and reliability of the instrument.

Darlene E. McCown, RN, PhD, PNP, FNP, currently serves as a family nurse practitioner in a shelter for homeless women and children. She is a faculty member at Saint John Fisher College, Rochester, New York, where she teaches family theory and clinical management in the graduate Family Nurse Practitioner Program. She is the author of more than 20 articles relating to children and families.

Judith R. Milligan-Hecox, RN, BSN, is Clinical Supervisor of the Visiting Nurse Association of Los Angeles, California, San Fernando Valley office. Her research interests include the study of coping strategies of patients with chronic obstructive pulmonary disease. Currently, she is a third-year MSN student at Azusa Pacific University.